A GLOSSARY OF
WORLD WAR II
MILITARY TERMS

A GLOSSARY OF WORLD WAR II MILITARY TERMS

Compiled from Official U.S. War Department and British War Office Documents

Nicholas G. Forte

TOWER & ANCHOR BOOKS
PENSACOLA, FLORIDA

ISBN: 979-8-9893207-0-7 (hardcover)
ISBN: 979-8-9893207-1-4 (paperback)

Library of Congress Control Number: 2023948152

Front cover image courtesy of U.S. Army Center of Military History

Published by Tower & Anchor Books
100 W Brainerd St.
Pensacola, FL 32501

www.towerandanchor.com

TABLE OF CONTENTS

FORWARD

World War II was a defining global conflict of the 20th century and brought with it a myriad of technical military terms that were essential for effective communication and coordination. *A Glossary of World War II Military Terms* aims to provide a concise compilation of more than 2,000 such terms, extracted from official U.S. War Department and British War documents of World War II.

The terms chosen generally deal with objects or practices of general military interest. Terms dealing with highly specialized activities have been excluded or defined to give only a general idea to non-specialized readers. As the list of terms is drawn from official literature, slang and colloquial terms, such as *Jeep* and *bazooka,* which were in wide use but unofficial, are not included.

In cross-referencing, a preference between two or more terms that had identical meanings has been established on the frequency of usage or the comparative standing of the source material. When two or more terms have identical meanings and one has a clear preference only the preferred term has been used. Other terms are merely listed and referred to the preferred term. When no clear preference can be established, all terms are defined in parallel language and cross-referenced.

Due to the relative extensiveness of the source material, most of the terms and definitions in this glossary are from American documents. Unless otherwise marked, the terms and definitions of this glossary were drawn from U.S. War Department sources. Nonetheless, British terms and definitions are included whenever possible, listed with both the equivalent American term and separately with a cross-reference to the American definition.

The notation *"British equivalent*: Same" indicates that the British Army also used the term. However, the lack of such notation does not necessarily mean the British did not use the term. The notation *"British equivalent*: no equivalent" does not necessarily imply that the function is not performed in the British armed forces, but rather that the British method was different enough from American methods to preclude giving a single term as even an approximate equivalent.

In addition to the glossary proper, multiple diagrams and tables are included to provide the reader with further resources to understand the meanings of several of the terms and to show the equivalence of terms between the U.S. and British armies. Tables showing the relative military ranks of all the major powers in World War II are also included.

For those intrigued by the intricacies of military history or engaged in scholarly pursuits, *A Glossary of World War II Military Terms* promises to be a valuable asset. As you explore the technical terminology within, take a moment to reflect on the context in which these terms were employed and the pivotal role they played in shaping the course of history. Through these words, we gain insight into the operational challenges, innovations, and strategies that defined an era of conflict and transformation.

NICHOLAS G. FORTE

MILITARY TERMS

Abatis.—(ab a TEE *or* AB a tis). An obstacle turned toward the enemy made of cut-down or fallen trees, or of small trees or saplings bent down, often interlaced with barbed wire.

Absent without leave.—Absent from post of duty without permission from proper authority and without intention of deserting. *British equivalent*: ***Absence from duty*** or ***illegal absence***.

Accessories.—Tools and equipment used in assembling, disassembling, cleaning, and protecting military apparatus, especially guns and vehicles.

Accompanying.—Attached to, and moving with, an attacking force. *British equivalent*: ***"in support of"*** *(or* ***"under command"***).

   Accompanying artillery.—Single batteries, platoons, or pieces attached to assault infantry regiments or battalions for their close support.

   Accompanying fire.—A rolling barrage or other fire delivered by accompanying artillery.

   Accompanying guns (or batteries).—Guns of an accompanying artillery unit.

   Accompanying tanks.—Single tanks attached to attacking units.

Accouterments.—A soldier's equipment, with the exception of his weapons and clothing. *British equivalent*: Same.

Acting.—Serving temporarily in another rank or capacity. For example, privates or privates, first class are sometimes designated acting corporals, with the authority but not the pay or permanent rank of corporal

Action.—An engagement or battle, usually one on a small scale. *British equivalent*: Same.

Action station.—An assigned position to be taken by an individual in case of an air attack.

Activate.—To organize a unit on the active list of the Army by assigning to it personnel and equipment with which it can operate as a distinct unit. Activate differs from ***constitute***, which establishes a new unit on the active or inactive list of the Army but does not give it physical existence.

Active defense.—Resistance by the use of weapons. Active defense does not include camouflage, scattering of vehicles, etc., which are called ***passive defense***.

A GLOSSARY OF WORLD WAR II MILITARY TERMS

Active air defense.—A defensive action taken to destroy attacking enemy aircraft in the air. Active air defense includes such measures as the use of fighter airplanes, antiaircraft artillery, and barrage balloons; it does not include the use of cover, concealment, and dispersion to guard against enemy air attack, which are employed as ***passive air defense.*** *British equivalent:* ***Active air defence.***

Adjust.—To correct the elevation and deflection of a gun so that its projectile will hit the target.

Adjustment fire.—Gunfire directed for the purpose of obtaining data for the correction of gunfire. Adjustment fire is often contrasted with ***fire for effect***, which is fire to destroy enemy targets or to accomplish some other tactical purposes.

Adjutant.—A staff officer of a command who is responsible for all official correspondence except combat orders, for the distribution of orders, and for other administrative duties. The adjutant, first staff section, of brigades and lower units is referred to as S-1.

Adjutant-general.—The adjutant of a division or larger unit, or of a service command.

Addressee.—The person or office to which a message is to be delivered.

Administration.—When unqualified, administration includes all phases of military operations not involved in the terms "tactics" and "strategy." It comprises supply, evacuation, sanitation, construction, maintenance, replacements, transportation, traffic control, salvage, graves registration, burials, computations pertaining to movements, personnel management, quartering, military government, martial law, censorship, and other allied subjects. *British equivalent:* Same.

Administration order.—*British term:* see ***Administrative order.***

Administrative map.—A map on which is recorded graphically information pertaining to administrative matters; such as supply and evacuation installations, train bivouacs, rear echelon, straggler line, collecting points for stragglers and prisoners of war, main supply road(s), and the line forward of which no lights will be shown; necessary tactical details also shown. *British equivalent:* Same.

Administrative order.—An order covering administrative details, such as traffic, supply, and evacuation, when the instructions are too voluminous to be included in paragraph 4 of the field order, and at other times when necessary to publish administrative instructions to the command; usually issued by divisions and higher units.

See also **Combat orders**. *British equivalent:* **Administration order** (often issued as an appendix to the divisional operation order).

Administrative plan.—A plan proposed for handling the traffic, supply, evacuation, and other administrative details of operations of a unit.

Administrative services.—Branches of the Army Service Forces primarily in charge of Army administration, such as the Judge Advocate General's Department or the Adjutant General's Department.

Admiral.—An officer in the Navy who has a rank equivalent to that of a general in the Army. A vice-admiral is equivalent to a lieutenant general; a rear admiral to a major general.

Advance.—The progress of a command toward the enemy. To move forward. To make progress in the direction of the enemy. *British equivalent:* Same (terminates upon contact with the enemy).

Advance base.—*a.* A place located toward the front from which supplies of ammunition or other matériel are issued to units in field operations. *British equivalent:* **Advanced base.**
    b. A forward area where depots for supplies of ammunition and other matériel may be located. *British equivalent:* **Advanced base.**

Advance by bounds.—An advance controlled by the assignment of successive movement objectives usually from one terrain line to the next. *British equivalent:* Same.

Advance by echelon.—An advance of a unit by successive movements of its component elements. *British equivalent:* None.

Advance by rushes.—A move forward in short, quick spurts. An advance by rushes is usually made by individual soldiers or small infantry units in the face of active, enemy, small-arms fire.

Advance command post.—A designated point forward from the regular command post and convenient to the commander or small staff party, or both, for the exercise of temporary control. *British equivalent:* **Command post** or **battalion headquarters** or **tactical headquarters**.

Advanced airdrome.—A temporary airport located near the front. An advanced airdrome usually has only limited repair facilities, and its landing strips, shelters, and buildings are not highly developed.

Advanced base.—*British term: a.* A locality in which are situated the advanced depots of ammunition, supplies, animals, and materials.
    b. When a force is maintained partly from outside the actual theater of operations, that part of the base inside the theater.

c. In combined operations, a base sufficiently close to the zone of operations to permit of supplies, etc., being sent direct from it to that zone. It may also be used for storing supplies, concentrating reinforcements, and establishing hospitals, rest camps, etc.

Advanced depot.—A supply point in the forward part of the communications zone in a theater of operations, ahead of the intermediate and base depots.

Advanced dressing station.—*British term*: A divisional dressing station formed by one or more companies of a field ambulance or light field ambulance where casualties are given first aid, sorted, and labeled, and from which they are evacuated. See ***Collecting station.***

Advanced HQ.—*British term*: see ***Combat post.***

Advanced surgical centre.—*British term*: A corps medical unit formed by a field transfusion unit and one or more field surgical units. Casualties requiring urgent surgical treatment in the field are evacuated to this centre from the ***advanced dressing station.***

Advance guard.—A detachment sent ahead of the main force to protect it against surprise and to facilitate its advance by removing obstacles, repairing roads and bridges, etc. The advance guard also locates the enemy, especially his main forces; and delays the enemy's advance long enough to permit the main force to prepare and deploy for action. *British equivalent*: ***Advanced guard.***

Advance guard action.—An attack or defense by an advance guard to occupy or deceive the enemy and to protect the advance and deployment of the main force.

Advance guard point.—The first unit of an advance guard support. The advance guard point consists of a small group of soldiers sent out ahead of the advance party to warn against surprise attacks.

Advance guard reserve.—The second of two main parts of an advance guard. It protects the main force and is itself protected by the advance guard support. Smaller advance guards do not have reserves.

Advance guard support.—The first of the two main parts of an advance guard. It is made up of three smaller parts, in order from front to rear, the point, the advance party, and the support proper. The advance guard support protects the second main part, the advance guard reserve.

Advance landing field.—A temporary landing field located near the front and provided with only the very necessary supplies and equipment for repair and servicing of aircraft.

Advance message center.—A communication center for the reception and relay of messages to facilitate communications with advanced units or units operating on a flank. *British equivalent*: **Advanced signal centre** or **report centre**.

Advance on.—Advance toward. *British equivalent*: Same.

Advance party.—*a*. A detachment that it sent out by and moves ahead of the advance guard support, following the advance guard point and going ahead of the advance guard support. The advance party protects the support and is itself protected by the advance guard point which precedes it. *British equivalent*: **Van guard**.

b. A detachment which precedes its unit to make administrative or other arrangements.

Advance post.—A concealed observation place well ahead of the main body of troops. An advance post is used for observing, listening, or otherwise securing information, and is linked with headquarters by some system of rapid communication.

Advance section.—The forward or most advanced subdivision of the communications zone. *British equivalent*: None.

Advancing fire.—See **Assault fire**.

Aerial mine.—A large, thin-walled container with a heavy charge of high explosives, used to destroy enemy surface defenses and installations. An aerial mine can be fitted with a parachute to prevent penetration and obtain the maximum blast effect.

Aerial observation.—Obtaining military information by observing or by taking photographs from aircraft for the purpose of liaison, reconnaissance, or directing artillery fire; aerial exploration of a near objective or area. Also called **air observation**. *British equivalent*: Same.

Aerial photograph.—A picture taken from any kind of aircraft. Aerial photograph, oblique (vertical); air photograph. See **Oblique (vertical) aerial photograph**. *British equivalent*: **Air photograph**.

Aerodrome.—*British term*: see **Airdrome**.

Aeronautical charts.—Maps upon which information pertaining to air navigation has been added; intended primarily for use in air navigation. They are classified as "sectional" (scale 1:500,000) and "regional" (scale: 1:1,000,000). *British equivalent*: **Aeronautical maps**.

Aerostatics.—The science or art of operating balloons and other lighter-than-air aircraft. Aerostatics is distinguished from *aviation*, which is the art of operating heaver-than-air aircraft.

Affirmation.—A solemn declaration made instead of taking an oath. If a person's religion forbids him to take an oath, he can make an affirmation which will have the same force of law as an oath.

Agency of signal communication.—A term embracing the personnel and equipment necessary to operate message centers, signal intelligence, signal supply, and messenger, pigeon, radio, visual, sound, and wire communication. *British equivalent*: **Signal unit.**

Aide.—See **Aide-de-camp.**

Aide-de-camp.—A member of the personal staff of an officer, usually a general officer in a high command; aide. His duties include receiving and transmitting orders or performing any other duties the general officer may assign him.

Aid man.—A man from the Medical Department attached to a company, battery, troop, etc., to give first aid to the wounded and to carry necessary information to the battalion or regimental surgeon.

Aid station.—An establishment of the Medical Department provided for the emergency treatment, sorting, and further disposition of casualties in combat. The first station on the route of evacuation to which the wounded are brought. Aid stations are usually established for each battalion in combat by the battalion medical detachment. *British equivalent*: **Regimental aid post** (in the case of a battalion or a similar unit).

Aigulette.—(AY gwa LET). A decoration of braided cord for the dress uniform of aides, military attachés, and officers of the General Staff Corps.

Air area.—An area assigned as a means of coordinating the air reconnaissance activities of various units having organic or attached observation aviation. *British equivalent*: **Air reconnaissance area.**

Air alert.—*a*. A position of aircraft kept in the air ready for immediate action.

b. A signal to take stations for an air alert.

Air alert method.—A method of air defense in which fighter airplanes are kept in the air, overhead or nearby, ready for immediate action. The air alert method is one of three methods of using fighter aviation in air defense; the other methods are the **ground alert method** and the **search patrol method.**

Air area.—A limited region within which a commander has responsibility for observation and reconnaissance from the air. The area is set up to enable various ground units having air observers to cooperate efficiently.

Air attack.—The attack of objectives on the earth's surface by aircraft.

Air base.—A command which is equipped and organized for sustaining the operations of one or more tactical air units. An air base consists of the personnel, airdromes, and other facilities necessary to support the operations of tactical air units currently using the facility. *British equivalent:* Same.

Airborne infantry.—Infantry units, including parachute troops, especially trained and equipped for transportation by air.

Airborne troops.—A general term used to include both parachute and air landing troops. *British equivalent:* Same.

Air Corps.—The name that was replaced on May 1, 1942, by U.S. Army Air Forces except by special reference. Officers were still commissioned in the Air Corps and Air Corps Reserve.

Aircraft signal.—A signal from an aircraft as a means of communication between air and ground forces when radio can not be used; airplane signal. An aircraft signal may be made by dipping the wings, dropping a flare, or by firing a type of fireworks giving off an intense colored light.

Aircraftman.—*British term:* A rank in the Royal Air Force equivalent to that of private or trooper in the Army. It is divided into leading aircraftman; aircraftman, 1st class; and aircraftman, 2nd class.

Aircraft warning service.—A warning system consisting of observers, information centers, and signal communication established by territorial commanders for the primary purpose of determining courses of hostile aircraft and of distributing information to industrial centers and to military and naval commands. *British equivalent: Air raid warning system.*

Air defense.—Defense against attack from the air; all measures used to prevent enemy air action or to reduce its effect; antiaircraft defense. *Active air defense* is direct defensive action taken to destroy enemy aircraft in the air by fighter aircraft, antiaircraft artillery, and barrage balloons. Active air defense over a large area is called *general air defense*; that over a small area or single objective is called *local air defense*. *Passive air defense* includes all means used on the ground to prevent or guard against air raids, such as the use of aircraft observers, air-raid wardens, blackouts, alerts,

13

shelters, and also the use of cover, concealment, and dispersion. *British equivalent*: **Air defence.**

Air defense area.—A territory that includes the objectives of a probable air campaign of the enemy, and for which protection must be provided. The United States was divided into air defense areas. For military purposes, each of these air defense areas was controlled by an interceptor commander and was further divided into **air defense regions.** For civilian purposes, each air defense area was divided into **warning districts.**

Air defense command.—An organization for the coordination of all measures of defense against enemy air operations, including aircraft warning services, pursuit aviation, antiaircraft artillery, balloon barrages, and passive antiaircraft defense measures. *British equivalent:* **Fighter command.**

Air defense region.—A division of an air defense area to which are assigned one group of interceptor planes and at least one signal warning regiment.

Airdrome.—A landing field, with the necessary additional installations for servicing, arming, operating, and maintaining military aviation units. Also called **airport.** *British equivalent:* **Aerodrome.**

Air echelon.—In combined ground and air operations, the air unit as distinguished from the ground forces which it supports.

Air force.—*a.* The branch of a nation's armed services that conducts military operations in the air.

b. The largest tactical and administrative unit of the U.S. Army Air Forces. An air force normally consists of a headquarters, a bomber command, an interceptor command, and air support command, and such other tactical air units and ground units as may be attached or assigned. Its duties include air defense, bombardment, ground-air support, and base services. An air force is the equivalent of a field army.

c. The **Air Forces** means the Army Air Forces.

Air Force Act.—*British term:* see **Articles of War.**

Air-ground liaison panel.—A large cloth strip used by ground troops for visual signaling to supplement radio communications with aircraft. Also called **air-ground panel.**

Air-ground net.—A system of radio communication linking aircraft in flight with ground stations.

Air-ground observer.—See **Air guard.**

Air-ground panel.—See **Air-ground liaison panel.**

Air guard.—A person posted to warn of the approach of enemy aircraft; air scout; air sentinel, air guard observer; antiaircraft lookout. Air guards are posted by units on the march or in any position open to air attack even though the area has a permanent warning system.

Air guard observer.—See **Air guard.**

Air-landing troops.—Troops carried in powered aircraft, or in gliders towed behind aircraft, who disembark after the aircraft or glider reaches the ground. *British equivalent:* Same.

Air liaison officer.—*British term:* An army officer attached to an RAF unit or formation, whose duties include the briefing and interrogation of pilots undertaking reconnaissance, and assisting the commanders of squadrons carrying out any Army tasks in the briefing an interrogation of air crews. He is responsible for passing on the results of the interrogation and for keeping pilots fully informed about the military situation. He also carries out a quick scrutiny of all air photographs received.

Airman (or airmen).—*British term:* In the Royal Air Force, includes warrant officers, non-commissioned officers, aircraftmen, apprentices, and boy entrants. See also **Enlisted man, Other ranks**, and **Ratings.**

Air observation.—See **Aerial observation.**

Air officer.—See **Air support officer.**

Air observation.—Obtaining military information by observing or by taking photographs from aircraft for the purpose of liaison, reconnaissance, or directing artillery fire; aerial exploration of a near objective or area. Also called *aerial observation. British equivalent:* Same.

Air operations officer.—An administrative and tactical officer, generally assigned to air fields and air units as the officer in charge of plans, policies, and operations, especially air-landing operations.

Airplane defense area.—The area beyond the range of friendly antiaircraft artillery. An airplane defense area is protected by fighter aircraft, assisted by night searchlights.

Airport.—A tract of land or water where aircraft can land or take off, and where facilities for their shelter, supply, and repair are available. An airport is often used for regular receiving or discharging passengers and cargo traveling by air. A military airport is usually called an **airdrome.**

Air photograph.—See **Aerial photograph.**

Airportable.—*British term*: Term applicable to the equipment, modified as necessary, that accompanies airtransported troops.

Air raid warning system.—*British term*: see **Aircraft warning service.**

Air reconnaissance.—The getting of military information by observation and taking pictures from aircraft; aerial exploration of a distant objective of area. Air reconnaissance is usually made over territory held by the enemy and is made to get information about military objectives in enemy territory, and the location, arrangement, and movement of enemy forces.

Air reconnaissance (air recce).—*British usage*: Reconnaissance, either visual or photographic, carried out from aircraft. There are four types:

    Strategic (strat R).—The search, usually by high-level photography of distant areas, for information that may influence the general course of the campaign.

    Tactical (tac R).—The search, by visual observation or photography, for information that my have an immediate effect on operations.

    Artillery (arty R).—Reconnaissance for artillery targets, and observation and control of artillery fire.

    Contact (con R).—Reconnaissance to locate the position of our own forward troops.

Air reconnaissance and observation.—The gaining of information through visual and photographic means carried in aircraft.

Air scouts.—See **Air guard.**

Air sentinel.—See **Air guard.**

Airship.—Any aircraft that is lighter than air and self-propelled by its own motors. There are three main types: non-rigid, semi-rigid, and rigid.

Airplane signal.—See **Aircraft signal.**

Air striking force.—In large commands, one or more groups of bombardment aviation, of the same or different classes, put under a single command.

Air superiority.—An advantage held by military aviation over enemy aircraft that permits air or ground operations in a locality without effective enemy air opposition. *British equivalent*: **Air superiority.**

Air Support Command.—A tactical and administrative unit of military aviation, organized, equipped, and trained to render air support to large ground units, usually an army.

Air support control.—An aviation unit located at the headquarters of the army or other large ground unit which it supports. The air support control manages the operations of the air support and maintains communications with the air units.

Air support missions.—Missions assigned air support aviation include both the immediate support of ground forces where contact with the enemy is imminent or has already been established, and the destruction or neutralization of timely but more distant targets to prevent or impede hostile movement, intervention or entry into combat.

Air support officer.—An officer in the Army Air Forces attached to a ground tactical unit to advise on air matters; air adviser; air officer. He represents the air support control. He also transmits communications between the commander of the ground unit supported, the air support control, and aircraft in flight.

Air task force.A group of air, ground, and service units needed to carry out air missions as outlined in a plan of military operations; task air force.

Air transport.—Aircraft used to move supplies, equipment, personnel, or vehicles.

Air transportation.—The moving of supplies, equipment, personnel, or vehicles by air, usually in transport airplanes or in gliders.

Airtransported.—*British term*: Term applicable only to troops who do NOT form part of airborne divisions but who may be transported by air for a special purpose.

Air warning net.—A system of radio stations set up to warn against attack from the air.

Alarm.—A warning, by bugle, cannon fire, siren, or other means, to summon military forces to meet an emergency.

Alert.—*a.* Readiness for action, defense, or protection.

b. To get ready for action.

c. A warning signal or a real or threatened danger, such as an air attack.

d. A period of time during which troops stand by in response to an alarm.

e. A condition of aircraft manned, armed, and ready to carry out a mission.

Alert station.—A position taken up by defensive aircraft between expected enemy aircraft and the objective to be defended.

Alignment.—*British term*: see **Alinement.**

Alinement.—A line upon which several elements are formed or are to be formed, or the dressing of several elements upon a line. *British equivalent:* **Alignment.**

All-around traverse.—Turning or swinging a gun on a mount that permits a gun to be turned in a complete circle in a horizontal plane. A gun has an all-around traverse when it can be turned, clamped, and fired in any horizontal direction without changing the position of the tripod or mount.

All-clear signal.—A prearranged signal to indicate that danger from enemy aircraft has passed.

Allotment.—Allocation of units for certain missions or assignment. Applicable to GHQ reserve tank units.

Alternate emplacement.—An emplacement prepared for occupation in case the principal emplacement becomes untenable or unsuitable. *British equivalent:* **Alternative site.**

Alternate firing position.—A firing position from which the same fire missions can be executed as from the primary firing position. Sometimes called **alternate position.** *British equivalent:* **Alternative position.**

Alternate rallying points.—A substitute location where troops assemble and reorganize after an attack when the first rallying point designated has become unsuitable.

Alternate traversing fire.—A method of covering a target that has both width and depth by firing a succession of traversing groups whose normal range dispersion will provide for distribution in depth.

Alternative position.—*British term:* see **Alternate firing position.**

Alternative site.—*British term:* see **Alternate emplacement.**

Amatol.—A high explosive made of a mixture of ammonium nitrate and TNT. It is used as a bursting charge in high explosive projectiles.

Ambulance loading post.—A location in the forward combat area where sick and wounded persons or animals are loaded into ambulances for transportation to a clearing station or hospital. *British* *equivalent*: **Car post.**

Ambulance station.—A point established for the administration and control of ambulance units and the regulation of movement of ambulances from front to rear and vice versa. *British equivalent*: None (performed at the **advanced dressing station**).

Ambush.—A concealed place or station where troops lie hidden for the purpose of attacking by surprise. Troops posted in such a position. To attack from such a position. *British equivalent*: Same.

Ammonal.—A high explosive substance, made of a mixture of ammonium nitrate, TNT, and flaked or powdered aluminum. Ammonal is used as a bursting charge in high explosive projectiles, and produces bright flashes on explosion.

Ammunition.—All materials used in discharging every kind of firearm or any weapon that throws projectiles. Ammunition includes power, shot, shrapnel, bullets, and cartridges, and also the means of igniting and exploding them, such as primers and fuzes. Chemicals, bombs, grenades, mines, and pyrotechnics are also ammunition.

Ammunition bearer.—A soldier who carries small-arms ammunition, including machine-gun belts and grenades, to supply ground troops in combat.

Ammunition belt.—*a.* A fabric or metal band with loops for cartridges that are fed from it into a machine gun or other automatic weapon. In this meaning, usually called **feed belt.**

b. A belt with loops or pockets for carrying cartridges or clips of cartridges. In this meaning, usually called **cartridge belt.**

Ammunition carrier.—*a.* A vehicle that accompanies guns and carries ammunition.

b. A member of a gun or mortar squad who carries ammunition and helps load in actual firing.

Ammunition point.—*a.* A place at the rear of a company or battery where ammunition is received from the ammunition supply point and from which it is issued to the company or battery.

b. British: A point at or in the immediate vicinity of which loaded **second line transport** ammunition vehicles are located, and from which the replenishment of forward ammunition echelons takes place. Also see **Distributing point.**

Ammunition railhead.—*British term*: A terminal point on the railway Line of Communications at which ammunition and explosive are transferred from rail to dumps of for forward transmission by road transport.

Ammunition supply point.—An advance point where ammunition is received, classified, stored, and issued to combat troops. The troops draw the major quantities of their ammunition from the ammunition supply point.

Anchor.—*a.* A device used to hold an object in place. Anchors are used to hold boats, vessels, mines, or floating bridges in place in the water. On land, a device that holds an object, such as a wire entanglement, in place on the ground is sometimes called an anchor.

    b. A key position in defense lines.

Annexes.—Anything added to a field order or other document to make it clearer or to give further details. Maps, photographs, schedules, etc., are annexes. Annexes, such as a medical annex, were frequently used to cover special activities or special units.. *British equivalent*: ***Appendices and traces*** or ***annexures.***

Annihilation fire.—See ***Counterpreparation.***

Antiaerial.—Opposed to enemy action that is directed from the air, such as descending parachute troops or gliders attempting to land troops.

Antiaircraft artillery intelligence service.—A system of observers and communication facilities established by antiaircraft artillery units for the purpose of gathering and transmitting information of enemy aircraft activities necessary for the proper employment of the antiaircraft artillery. *British equivalent*: ***Royal Observer Corps, RAF*** (searchlight units and spotters within the unit carry out these duties).

Antiaircraft defense.—That class of defense provided by the coordinated employment of air and ground forces against attack from the air. It includes passive means of defense. *British equivalent*: ***Antiaircraft defence*** (including ***passive air defence***).

Antiaircraft lookout.—See ***Air guard.***

Antimechanized defense.—All measures used to protect troops, installations, and establishments against mechanized, motorized, or armored units; ***antitank defense.*** Antimechanized defense employs such means and antitank guns and grenades, ditches, traps, and mine fields. *British equivalent*: ***Anti-tank defence.***

Antipersonnel bomb.—A bomb designed for use against individuals. An antipersonnel bomb is a small light bomb that bursts into fragments. See also ***Fragmentation bomb.*** *British equivalent*: ***Anti-per-***

sonnel bomb.

Antipersonnel mine.—A mine designed for use against individuals; personnel mine. *British equivalent*: ***Anti-personnel mine***.

Antitank defense.—See ***Antimechanized defense***.

Antitank ditch.—A ditch designed to stop the passage of track-laying vehicles. Sometimes called a **tank ditch**. *British equivalent*: ***Anti-tank ditch***.

Antitank grenade.—A high explosive grenade used against tanks or other armored vehicles; antitank rifle grenade. It is fired from a rifle by means of a special attachment. *British equivalent*: ***Anti-tank grenade***.

Antitank mine.—A device consisting of a container with a quantity of high explosive that detonates when pressure is exerted on it; also, any device similarly operated. *British equivalent*: ***Anti-tank mine***.

Antitank mine field.—A grouping of antitank mines placed in concealed position so spaced as to stop or impede the progress of track laying or wheeled vehicles. *British equivalent*: ***Anti-tank minefields***.

Antitank obstacle.An obstruction or barrier set up to stop or slow down enemy tanks or other armored vehicles; tank obstacle. Antitank obstacles include ditches, wire rolls, concrete pillars and blocks, etc. *British equivalent*: ***Anti-tank obstacle***.

Antitank rifle grenade.—See ***Antitank grenade***.

Antitank rifle grenadier.—A soldier trained in the use of grenade-throwing devices and tactics against armored vehicles.

Antitank rocket.—A rocket grenade used against tanks or other armored vehicles. It is fired from a special launching tube, which is held on the shoulder or on a prop when it is fired. The rocket attachment fires the grenade on the same principle as a pyrotechnic rocket is shot through the air.

Antitank weapons.—Those weapons whose primary mission is employment against armored vehicles. *British equivalent*: ***Anti-tank weapons***.

Appendices and traces.—*British term*: see ***Annexes***.

Application of fire.—Placing gunfire upon desired targets or upon a zone.

Appointing authority.—An officer who selects and appoints a board or court to investigate and decide a dispute, a question, or an action

21

between persons under his command. The exercise of this authority is a function of command and not rank. The appointing authority is also the **reviewing authority**.

Appreciation of the situation.—*British term*: see **Estimate of the situation**.

Approach.—A route by which a place or position can be approached by an attacking force. The route leading to anything, as a bridge. *British equivalent*: Same.

Approach formation.—A formation taken by a unit as it nears a combat area or contact with the enemy.

Approach march.—The advance, usually in extended dispositions from the point where hostile medium artillery fire is expected or air attack is encountered to the point of effective hostile small-arms fire. It ordinarily commences with the development of companies and larger units and terminates with their complete or partial deployment as skirmishers. *British equivalent*: Same.

Approach march formation.—An arrangement of advancing forces in an approach march. This formation has two stages: first, the advancing forces are broken up into small groups; second, the small groups are deployed in the combat area.

Approach trench.—A trench serving to connect fire trenches from front to rear. *British equivalent*: **Communication trench**.

Appropriation.—A sum of money set apart for some special use; money provided by a legislative body for a special purpose.

Approval authority.—See **Reviewing authority**.

Apron.—*a.* An area with a hard surface in front of a hanger or aircraft shelter.

b. A network that slants out from either or both sides of the center stakes of a barbed-wire entanglement.

c. A removable screen of camouflage material placed over, or in front of, artillery guns.

Arc of fire.—*British term*: see **Sector of fire**.

Area bombing.—Bombing a general area where there is no attempt to hit particular targets in the area. Area bombing differs from **precision bombing**, which bombing at a specific target, and from **pattern bombing**, which is systematic covering of a target according to a plan.

Area fire.—Fired directed to cover an entire area, either part at a time or all at once. Area fire differs from precision fire, which is directed against a definite part of a specific target in an area.

Area target.—A target for gunfire or bombing covering a considerable space, such as a munitions factory, airport, or freight yard. An area target differs from a *point target*, which is a particular object or structure.

Arm.—*a.* A weapon used in war. In this meaning, usually called arms.

　b. To supply with weapons.

　c. A branch in the Army whose main function is to engage directly in combat. The principal arms are Coast Artillery Corps, Field Artillery, Army Air Force, Cavalry, and Infantry. The arms are know collectively as the line of the Army. Arm differs from service, a branch of the Army whose chief mission is not combat but supply, administration, transportation, or medical care.

　d. To put a fuze in a bomb or projectile into proper condition to explode when it hits.

　e. British equivalent: Same (in all meanings).

Armed forces.—All military forces of a nation, including such organizations as the army, navy, marine corps, and coast guard; armed services

Armament.—*a.* The weapons of a particular vehicle, airplane, unit, etc.

　b. War equipment and supplies. Armament includes all weapons, ammunition, vessels, fortifications, and organized personnel, industry, and services used in land, sea, and air warfare.

Armament train.—The total group of vehicles used for transportation of guns, ammunition, and stores. An armament train also quarters the personnel connected with the group while traveling.

Armistice.—A temporary stop in fighting between two forces, armies, or nations by agreement of both sides. *British equivalent:* Same.

Armored car.—An armed and armored motor vehicle designed primarily for reconnaissance. *British equivalent*: *Armoured car.*

Armored combat.—Fighting carried on by forced of armored vehicles, usually supported by motorized infantry, artillery, and air units.

Armored Command.—The branch of the Army Ground Forces trained and equipped to fight with tanks and other mechanized vehicles. Prior to July 3, 1943, called *Armored Force.*

Armored force.—A combined force comprising reconnaissance, assault, and supporting troops of more than one arm or service, transported in wheeled or track-laying type motor vehicles, the bulk of which are provided either with partial or complete armor. *British equivalent:* **Armoured troops.**

Armored Force.—See **Armored Command.**

Armored vehicle.—A motor car or track-laying vehicle protected by steel plates, used for transporting, scouting, and combat. Armored vehicles include scout cars, armored cars, tanks, self-propelled artillery, and wheeled or half-track carriers. *British equivalent:* **Armoured vehicle.**

Armoured carrier.—*British term:* An armored motor vehicle for carrying men or material. See also **Loyd carrier** and **Universal carrier.**

Armoured command vehicle.—*British term:* see **Command car.**

Armoured fighting vehicle.—*British term:* A vehicle, either wheeled or tracked, in which the crew serves the armament with which it is equipped from behind bullet-proof plate, and which is designed to allow its weapons being fired while it is on the move.

Armorer.—A man who repairs and services small arms, fills ammunition belts, and performs similar duties necessary to keep small arms ready for use.

Armorer-artificer.—A person in certain types of units who makes minor repairs on weapons, does simple carpentry work, and aids in checking and distributing all supplies except rations or water.

Arms of the Service.—*British term:* All branches, taken collectively, of the Army. The combatant branches are called "The Arms; the administrative branches, "The Services."

Army.—*a.* The largest administrative and tactical unit of a land force, made up of a number of corps and divisions; field army.

b. All the military forces of a nation, exclusive of the naval forces and, in some countries, exclusive of the air forces.

c. **Shock Army.**—A Soviet army composed of picked tank and mechanized corps and rifle divisions. These armies normally were retained in the reserve of the high command.

Army Act.—*British term:* see **Articles of War.**

Army Air Forces.—One of the three major subdivisions of the U.S. Army, comprising all the aviation of the Army, together with its personnel, equipment, supply, etc.; Air Forces. Prior to May 1, 1942, called **Air Corps.** The other two subdivisions are the **Army Ground Forces** and the **Army Service Forces.**

Army air support.—*British term*: Term which, in its widest sense, includes the provision of fighter protection, air reconnaissance, attack on ground targets, and all forms of air transport. Air support may be:

Direct.—Air intervention against enemy land forces actually engaged in the battle.

Indirect.—Air action having a general influence on the course of the campaign.

Requests for air support are passed through air support signal units, which proved the necessary ***tentacles*** to Army and RAF formations.

Army corps.In many armies, this is the name of a tactical unit larger than a division and smaller than an army. An army corps usually consists of two or more divisions together with auxiliary arms and services. In both the U. S. Army and the British Army this name has been shortened to simply ***Corps***.

Army depot.—A supply point, located within the area of an army and designated by the army commander, where supplies from the communications zone or from local sources are received, classified, stored, and distributed.

Army engineer.—A senior engineer officer in charge of engineer troops assigned to an army.

Army exchange.—A military organization that sells merchandise and services to military personnel and other authorized personnel. Often called an ***Exchange*** or ***post exchange***.

Army Ground Forces.—One of three major subdivisions of the U.S. Army, comprising al the ground combat branches of the Army, with their personnel, equipment, supplies, etc.; Ground Forces. The other two subdivisions are the ***Army Air Forces*** and the ***Army Service Forces***.

Army group.—See ***Group of armies*** and ***Front***.

Army of occupation.—An army established in conquered territory to maintain order and to insure the carrying out of peace or armistice terms.

Army of the United States.—A temporary military organization of the United States established in time of war or national emergency. It includes the Regular Army, the National Guard of the United States, the National Guard while in federal service, the Organized Reserves, all Selected Service personnel, and all officers who are appointed in the Army of the United States but not in any particular component. It differs from the **United States Army**, which includes only the permanent military forces such as the Regular Army, the National Guard of the United States, and the Organized Reserves.

Army quartermaster depot.—A depot which is located as far to the front as practicable but out of hostile artillery range. They are established for the reception and temporary storage of supplies which because of the situation must be kept closer to the army units than the advance section of the communications zone.

Army Register.—An annual publication issued by the Adjutant General's Office that gives the names of Regular Army officers, active and retired, brief summaries of the training and service of each, and similar information.

Army Regulations.—The officially printed announcements, usually in pamphlet form, of current War Department policies and rules. *British equivalent*: **King's Regulations**.

Army service area.—An area at the rear of a combat zone where the administrative establishments and service troops for an army are located.

Army Service Forces.—One of the three major subdivisions of the U.S. Army. It provides general administration, transportation, supply, evacuation, and other services to meet the requirements of the Army. The other two subdivisions are the **Army Air Forces** and the **Army Ground Forces**. *British equivalent*: **Royal Army Service Corps** and **Royal Army Ordnance Corps**.

Army tank.—*British term*: see **Infantry tank**.

Arrive.—To reach a designated point or line. Refers to the head of a unit. *British equivalent*: Same.

Arsenal.—A building or station for manufacturing, repairing, storing, or issuing weapons and ammunition.

Articles of War.—A code of laws governing the conduct of all persons n the Army or subject to military law. The Articles define military offenses, prescribe the composition and procedure of courts-martial, and fix the punishment of each crime. The corresponding code of the United States Navy is called "Articles for the Government of the United States Navy." *British Equivalent*: **Army Act** (for the Army), **Air Force Act** (for the Royal Air Force), and **Navy Discipline Act** (for the Royal Navy).

Artificer.—A soldier skilled with his hands in one kind of work, such as carpentry, painting, blacksmithing, etc.

Artificial obstacles.—Obstacles prepared by human agency; they may be fixed or portable. *British equivalent:* Same.

Artillery in support of (or "under command").—*British term: see* Accompanying artillery.

Artillery officer.—*a.* The senior officer in the artillery section of a large unit. He is the adviser to the commander and staff on all artillery matters.

    b. An officer of the artillery branches of the service.

Artillery assigned to corps.—*British term: see* **Artillery with the corps.**

Artillery mil.—An American unit of angle measurement equal to 1/6400 of a circle. Scales of sighting instruments are shown in artillery mils. An artillery mil is slightly smaller than an infantry mil. 100 artillery mils =98.2 infantry mils.

Artillery plan.—A plan that is worked out for the use, coordination, and control of artillery units in a particular combat operation.

Artillery position.—A position selected for and occupied by an artillery fire unit for the delivery of fire. *British equivalent*: Same (usually spoken as battery position *or* troop position).

Artillery preparation.—Intensive artillery fire delivered on hostile forward elements (short preparation) and other objectives (longer preparation) during the period immediately prior to the advance of the infantry from its line of departure to attack.

Artillery survey.—The observation and mapping of a ground area for the purpose of finding the locations of targets and the positions from which the most effective fire can be delivered on them.

Artillery with the corps.—A term used to indicate all the artillery in a *corps;* includes corps, division, and attached artillery. *British equivalent*: **Artillery assigned to corps.**

Assault.—To close with the enemy in order to employ weapons and shock action. When delivered by mounted troops, it is called the "charge". To deliver a concentrated attack from a short distance. To close with the enemy in hand-to-hand combat. *British equivalent*: Same.

Assault boat.—A small boat used to carry troops and weapons across a river, in order to make an attack; storm boat. If silence is required in making the assault, assault boats propelled by paddles are used; if speed is required, assault boats propelled by outboard motors are used.

Assault echelon.—One or more units of an attacking force used in close combat to begin and lead the attack; **assault wave.**

Assault fire.—Fire delivered by attacking troops as they close on an enemy to engage him at close range or in hand-to-hand fighting. Also known as **advancing fire.** Assault fire is usually delivered by troops advancing at a walk.

Assault, general.—An assault delivered on an extended front under coordination of a higher commander. *British equivalent:* **General attack.**

Assault gun.—Any of various sizes and types of guns that are self-propelled or mounted on tanks and are used for direct fire from close range against point targets.

Assault, local.—An assault initiated and executed by a small unit (squad, section, platoon, company, battalion) in order to take immediate advantage of local conditions. *British equivalent:* **Local attack.**

Assault wave.—See **Assault echelon.**

Assemble.—*a.* To gather or come together in close formation.

b. A command for units of troops to come together in close formation.

c. To put together the parts of a gun or other mechanism.

Assembly.—*a.* The regular grouping, in close order, of the elements of a command. *British equivalent:* **Forming up** or **parade**

b. The grouping of units in areas, prior to or following combat, for the purpose of coordination or reorganization preceding further effort or movement. *British equivalent:* **Forming up** or **concentration.**

Assembly area (or position).—The area in which elements of a command are organized preparatory to further action. For example, in the attack, liaison with supporting arms is arranged: objectives and other missions are assigned to component units.

Assembly position.—*British usage:* When used generally, this term applies to where troops assemble for any particular purpose. In certain circumstances it has a more definite significance, i.e.:

(i) **Night operations.**—The point at which the normal march formation will be abandoned and battle formation adopted.

(ii) **River crossing.**—A point where the assaulting troops assemble. It should be well away from the water obstacle and protected from ground and air observation.

Assembly trench.—A trench in which troops are organized preparatory to action.

Assessment of value of intelligence report.—*British term*: see **Evaluation of information.**

Assign.—To place an individual, unit, or item of equipment in a military organization so that the individual, unit, or item so placed becomes an organic part of the organization.

Attach.—To place an individual or unit temporarily in a military organization for duty, rations, or quarters without making the individual or unit an organic part of the organization. *British equivalent*: Same.

Attaché.—(AT a SHAY). A person serving on the official staff of an ambassador or minister to a foreign country. A military attaché is a military observer who reports to his own government on the military plans and developments of the country in which he is stationed.

Attached unit.—A unit placed temporarily under the direct orders of the commander of another unit to which it does not organically belong. *British equivalent*: **"under command."**

Attack.—An advance upon the enemy to drive him from his position.

Attack, continuing.—An aggressive action continued after an objective has been reached in order to prevent the enemy from reconstituting his defense on a rearward position. See also **Exploitation.** *British equivalent*: **Exploitation.**

Attack (or Attacking) echelon.—The leading echelon in attack. For example, in attack, it comprises infantry units which are advancing by fire and movement to close with the enemy. *British equivalent: Leading troops in attack.*

Attack position.—The last covered or concealed position taken by an attacking force before moving off to the line from which the attack is launched.

Attack wave.—The part of a combat unit that carries out the actual attack against an enemy. Generally each attack wave is followed and supported by another attack wave.

Authentication.—Proof by proper signature or seal that a military order, paper, or record is genuine and official; proof by secret signal that a message is genuine.

Autogyro.—A heavier-than-air aircraft, able to take off rapidly, climb steeply, and land with almost no run; gyroplane. An autogyro is supported in flight largely by the action of air on a set of freely revolving planes or blades, rather than by fixed wings as in an airplane. It is moved forward by an ordinary motor-driven propeller.

Autoloading.—Loading itself. An autoloading gun has a mechanism that throws out the used shell, puts in a new one, and prepares the gun to be fired, but does not fire it.

Automatic.—*a.* Self-acting; moving or acting by itself. An automatic gun throws out the used shell, puts in a new one, and continues to fire until the pressure on the trigger is released. Automatic means completely self-acting as distinguished from semiautomatic, which means partly self-acting.

   b. An automatic pistol.

Automatic fire.—Continuous fire from an automatic gun until pressure on the trigger is released. Automatic fire differs from semiautomatic fire, which requires a separate trigger pull for each shot fired.

Automatic firearm.—See ***Automatic gun.***

Automatic gun.—A firearm, gun, or cannon that fires continuously until the pressure on the trigger is released; automatic firearm; automatic weapon.

Automatic pistol.—A pistol that has a mechanism that throws out the empty shell, puts in a new one, and prepares the pistol to be fired again.

Automatic rifle.—A rifle that has a mechanism that throws out the empty shell, puts in a new one, and prepares the rifle to be fired again. Some models are capable of either semiautomatic fire or full automatic fire.

Authorized allowances for equipment.—The quantity of items authorized for issue to a unit or organization in accordance with the Tables of Allowances, Tables of Basic Allowances, Tables of Equipment, Tables of Organization and Equipment, or special authorizations.

Automatic supply.—A process of supply under which deliveries of specific kinds and quantities of supplies are moved in accordance with a predetermined schedule. ***Daily automatic supply*** means that supplies are dispatched daily to an organization or installation. *British equivalent:* ***Normal supply.***

Auxiliary.—*a.* A person or organization that assists or supports a military unit in carrying out a duty or task.

b. Assisting or carrying out a duty or task.

c. Member of the Women's Army Auxiliary Corps (WAAC) who has a rank equivalent to that of a private in the Army.

Auxiliary aiming point.—A point or object used for laying a gun on a target that cannot be seen. The gunner adjusts the gun so that when the sight is aimed at the auxiliary aiming point the gun is laid on the target.

Auxiliary airdrome.—See **Satellite field**.

Auxiliary target.—A point at a known distance from the actual target; registration target. An auxiliary target is used as an adjusting point before firing on the actual target. Fire is delivered and adjusted on the auxiliary target. When the adjustment is complete, the necessary correction is put on the gun to swing it over the actual target. Auxiliary targets are used when fire on the actual target is intended to surprise the enemy.

Auxiliary arm.—Any arm that assists the principal arm assigned the mission of gaining or holding ground. *British equivalent*: **Supporting arm**.

Aviation.—*a*. Aircraft and the personnel needed to operate, maintain, and repair them.

b. The art of flying in an airplane; navigation of heaver-than-air aircraft. In this meaning, aviation is distinguished from **aerostatics**, which is the art or science of operating lighter-than-air aircraft.

AWOL.—Absent without leave; absent from post of duty without permission from proper authority and without intention of deserting.

Axis of advance.—A line of advance, often on a road or group of roads or a designated series of locations, extending in the direction of the enemy. *British equivalent*: Same.

Axis of evacuation.—A route by which matériel and personnel may be sent to the rear.

Axis of movement.—A line along which troops move either toward the front or toward the rear.

Axis of signal communication.—The initial and probable successive locations of the command post of a unit, named in the direction of contemplated movement. The axis of signal communication is the main route along which messages are relayed or sent to and from combat units in the field. *British equivalent*: **Signal communication along centre line** (armoured) or **main axis of advance** (infantry).

Axis of supply.—A route by which supplies are brought forward.

Axis of supply and evacuation.—A route in a combat zone by which supplies are brought forward and matériel and personnel may be sent to the rear.

Azimuth.—An expression of horizontal direction as an angle from north. The azimuth of a point is the angle formed at the observer between a north-south line and a line from the observer to the point. It is measured in degrees or mils clockwise from north. The north-south line may be true north, magnetic north, or grid north. See also **Bearing**.

Balanced stocks.—The accumulation of supplies of all classes, and in quantities determined as necessary to meet requirements for a fixed period of time.

Ball ammunition.—Cartridges containing solid bullets. Ball ammunition is general-purpose small-arms ammunition.

Ball cartridge.—A projectile that consists of a cartridge case, a primer, powder, and a solid bullet. Ball cartridges are general-purpose small-arms ammunition for standard service.

Ballistics.—The science of the motion of projectiles. *British equivalent*: Same.

Balloon barrage.—A barrier of captive balloons, with or without connecting cables or supported nets, against which hostile airplanes may be expected to run or because of which they may be forced to fly high over an area that it is desired to defend. *British equivalent*: Same.

Ball turret.—A gun turret, shaped like a ball, on the lower or bottom part of a bomber. A ball turret is power-driven, and can swing its guns to deliver fire in any direction.

Band.—*a.* Two or more lines of wire entanglements or other obstacles, arranged one behind the other. Each line of obstacles is a **belt**.
b. A path of fire, usually from machine guns.
c. A particular range of wave lengths in radio broadcasting.

Band of fire.—Fire, usually from one or more automatic guns, that gives a cone of dispersion so dense that a man trying to cross the line of fire would probably be hit. It is grazing fire, or at least part of grazing fire. A band for fire may be used as a **final protective line**.

Bandoleer.—A cloth belt that is divided into pockets for holding cartridges or clips of cartridges for small arms. *British equivalent*: Same.

Bangalore torpedo.—A metal tube or pipe that is packed with a high explosive charge. A Bangalore torpedo is chiefly used to clear a path through barbed wire or mine fields.

BAR.—See **Browning automatic rifle.**

Barbed-wire entanglement.—An obstacle or barrier of barbed wire, put up to prevent or hinder an enemy from entering an area.

Barracks.—A building or group of buildings used as living quarters for soldier.s

Barrage.—a.Prearranged fire on a line or lines, either stationary or moving. A concentration of artillery or machine-gun fire used to close off part of the front from enemy assault, to isolate part of the enemy position and prevent its being reinforced, or to protect an infantry advance by a curtain of fire moving in front of it. Also called **barrage fire.** *British equivalent*: Same.

b. A protective screen of balloons that are moored to the ground and kept at given heights to prevent or hinder operations by enemy aircraft at low levels. Steel cables or nets often hang from the balloons to make it dangerous for aircraft to fly beneath them. In this meaning, also called **balloon barrage.**

Barricade.—To fortify or close with a barrier, usually applied to roads. *British equivalent*: Same. *British equivalent*: Same

Barrier.—A group of obstacles, either natural or artificial, or both, that block or restrict entrance into an area.

Barrier line.—A traffic control boundary beyond which vehicles may not pas until other traffic with priority has gone through.

Barrier tactics.—Tactics based on the use of barriers that are defended by artillery or machine-gun fire to prevent or hinder the advance of the enemy in an area.. *British equivalent*: None.

Base.—a. A unit or organization in a tactical operation around which a movement or maneuver is planned and performed. In this meaning, usually called base unit.

b. A station or installation from which a military force operates and from which supplies are obtained.

c. A line used in mapping, surveying, or fire control as a reference from which distances and angles are measured.

d. *British*: A subarea organized to include two or more depots of men, animals, or material. When a force is wholly maintained within the actual theater of operations, the term "base" takes the place of the term **advanced base.**

Base area.—*British term*: see **Base section.**

Base airdrome.—An airdrome in the communications zone of a theater of operations. It is similar to a permanent peacetime airport except that its buildings are of a more temporary nature.

Base command.—An area organized under one commander of military operations to maintain a military base or group of bases. A base command is usually smaller than a theater of operations, and it is established for the administration of all military operations connected with the base or bases operating within it.

Base depot.—A supply point in the rear part of a communications zone in a theater of operations, behind the advance and intermediate depots.

Base of fire.—The supporting weapons of the unit of the attacking echelon when emplaced in firing positions to support the advance. For example, a base of fire is said to be established when units of the attacking echelons have been placed in departure positions and supporting weapons occupy firing positions with assigned target areas or sectors of fire to support the attack. The purpose of the organization of a base of fire is to bring about close coordination between the advance of the attacking echelon and the fire of supporting weapons.

Base of operations.—An area from which a military force begins its offensive operations, to which it falls back in case of reverse, and in which supply depots are organized. The base of operations is in the communications zone of a theater of operations.

Base repair.—Repair work done at a station by maintenance men specially trained and equipped to make repairs which cannot be done by line maintenance crews and which do not require major operations.

Base reserves.—Supplies accumulated and stored in depots for the purpose of establishing a general reserve, under the control of the commander of the theater, for the theater of operations as a whole. *British equivalent*: Same.

Base section.—The rear area of subdivision of the communication area. *British equivalent*: **Base area.**

Base spray.—Fragments of a bursting shell that are thrown out to the rear in the line of flight in contrast with **nose spray,** which are fragments of a bursting shell that are thrown out to the front, and **side spray,** which are fragments of a bursting shell that are thrown to the side.

Base sub area.—*British term*: An area round a base port which contains base depots and installations required to handle the type of traffic to be imported through the port. The local administration of these depots and installations and of the port is the responsibility of the headquarters of the base sub-area.

Base unit (or base of movement).—The unit on which a movement is regulated; base; base element. *British equivalent*: None.

Basic load.—*a.* The prescribed allowance of ammunition, supplies, and equipment carried by a unit.

b. The load on a structural member or part of an aircraft in steady flight.

Basic tactical unit.—The fundamental unit capable of carrying out an independent tactical mission in any branch of the Army, such as a rifle battalion in Infantry, a battery in Artillery, and a flight in the Army Air Forces.

Basic training.—Elementary training in military techniques and tactics. Basic training includes general subjects in addition to elementary training in the particular arm or service to which an individual is assigned.

Batman.—*British term*: A soldier employed as an officer's servant or groom. Previously called soldier-servant. See ***Orderly***.

Battalion.—A tactical unit made up of a headquarters and two or more companies, batteries, or similar organizations. It may be part of a regiment, but separate battalions exist that are administrative as well as tactical units.

Battalion aid station.—An emergency medical station set up close to the front by battalion medical personnel. Here the wounded are given emergency treatment and returned to duty or evacuated to the rear.

Battalion combat train.—A unit, organically part of a battalion, that furnishes a movable reserve or ammunition and a means of transporting it to batteries or companies.

Battalion defense area.—See ***Defense area***. For example, a section of the principal or outpost zone of resistance assigned to a battalion by the regimental or higher commander. Troops are disposed in the battalion defense area in width and depth so as to provide all-around defense of the area and mutual support for adjacent defensive areas.

Battalion train.—The vehicles and personnel that provide a battalion with facilities for supply, maintenance, and evacuation. Battalion trains comprise the battalion section of the regimental transportation platoon and the battalion section of the regimental medical detachment.

Battery.—*a.* A tactical and administrative artillery unit corresponding to a company or similar unit in other branches of the Army.

b. A group of guns or other weapons, such as mortars, machine guns, artillery pieces, or searchlights, set up under one tactical commander in a certain area.

c. A gun *"in battery"* is a gun in firing position. A gun *"from battery"* is a gun not in firing position.

Battle casualty report.—A report to The Adjutant General listing by name and serial number all men killed, wounded, missing in action, or captured as a result of enemy action.

Battlefield recovery.—The removal from the battlefield by combat personnel (supplemented as necessary by service personnel) of disabled or abandoned matériel pertaining to both enemy and friendly troops, and its movement to defilade, to an axis of matériel evacuation, to a recovery collecting point, or to a maintenance or supply establishment where it can be returned to service immediately, or repaired and reissued.

Battle honor (or battle honors).—A colored streamer flown from the staff of the flag, color, guidon, or standard of a unit, or silver band fastened around the staff of the guidon of a unit. A battle honor is awarded or authorized by the War Department, or by foreign governments, as a decoration for meritorious service by a unit in combat, and shows the battles or campaigns in which it was earned.

Battle map.—*a.* A map, prepared normally by photogrammetric means and at a scale of 1:20,000, for the tactical and technical needs of all arms. *British equivalent*: None.

b. British term: see **Situation map**.

Battle position.—The position of principal resistance in defense, consisting of a system of mutually supporting defensive sectors (areas) disposed in breadth and depth. *British equivalent*: **Defensive position (or system)** or **point of manoeuvre** (on a small scale).

Battle reconnaissance.—The continual observation, made under combat conditions, of the terrain, disposition of the enemy, etc. It is made during or immediately before battle, when in close contact with the enemy.

Battle reserves.—Supplies gathered in the neighborhood of a battle-field, in addition to the reserves of the unit and of the individual soldiers.

Battle sight.—The rear sight on a rifle with a large aperture or notch set for a convenient range. A battle sight is used for emergency fire at close range when there is no time or need for accurate adjustment.

Bayonet.—A blade attached to the muzzle end of a rifle. It may be detached and used as a separate hand weapon.

Bayonet scabbard.—A carrying case for a bayonet, attached to a soldier's belt by two hooks.

Beach defense.—That part of the ground organization for defense against landing attacks which is located at or near the beach for resistance at the water's edge. *British equivalent*: Same.

Beachhead.—Position occupied by advance troops landing on a hostile shore to protect landing areas for other friendly troops and for supplies on the beach or at a port. *British equivalent*: Same.

Beach maintenance area.—*British term*: An area in the vicinity of the beaches comprising service units and detachments for the maintenance, in the initial stages, of troops landed in a combined operation.

Beach party.—Men detailed to assist in the landing of troops and supplies.

Beach reserves.—An accumulation of supplies of all classes established at dumps on the beach. *British equivalent*: Same.

Bearing.—*a.* An expression of the horizontal direction of a point from an observer it terms of an angle, measured in degrees, from a line extending north and south through the observing point. The north-south line may be either true or magnetic north. Since the angle is measured east or west from north or south, a bearing is never greater than 90 degrees. In military practice, expression of direction in terms of *azimuth* is preferred to bearing.

b. British: True bearing is the angle a line makes with the true north line. Magnetic (or grid) bearing is the angle a line makes with the magnetic (or grid) north line. In each case the angle is measured from North by East and South.

Beaten zone.—The pattern formed by the cone of fire when it strikes the ground. *British equivalent*: Same.

"B" echelon transport.—*British term*: see **Field train**.

Belt.—*a*. A cloth strip with loops, or a series of metal links with grips, for holding cartridges which are fed into an automatic gun.

b. A band of leather or webbing, worn around the waist and used as support for weapons, ammunition, etc.

c. A strip of terrain, usually parallel to the front.

d. A line of obstacles used for defense. A line of wire entanglements is a belt. Two or more belts form a **band**.

Belt-fed.—Supplied with cartridges from a feed belt for automatic weapons.

Belt-loading.—An arrangement of ammunition for an automatic, belt-fed gun in an ammunition belt.

Berm.—A shelf near the top of a trench or other dugout shelter. A berm prevents caving banks or sides from falling into the trench or shelter and provides a support for beams.

Billet.—*a*. A shelter that consists of private or nonmilitary public buildings.

b. To quarter troops in private or nonmilitary public buildings.

c. *British*: Personnel, animals, and transport accommodated in civil buildings are said to be billeted. Billets are usually adopted in civilized countries for forces not in close touch with the enemy. They provide concealment from air observation and give the best form of shelter and rest.

Bivouac.—*a*. An area in the field in which troops rest or assemble. A bivouac may have no overhead cover, or only natural cover, shelter tents, or improvised shelter. *British equivalent*: Same or **harbour** (for armoured formations or units) or **leaguer** (in desert warfare).

b. *British*: A camp without tents or huts, where troops make their own shelters. Its essential characteristic is readiness for action. Bivouacs enable the most favourable dispositions of troops to be made from the tactical point of view; but may be injurious to the health in cold and wet weather.

Black powder.—A mixture of powdered potassium nitrate or sodium nitrate, charcoal, and sulfur; gunpowder. Black powder is a low explosive used in igniters, primers, fuzes, and blank-fire charges. *British equivalent*: Same.

Blanket.—*a*. A layer of smoke, clouds, or fog covering troops or operations.

b. To lay a protective screen of smoke over, or in front of, friendly troops; to blind enemy troops with a cover of smoke.

c. To prevent a sound locator from discovering bombers by flying a pursuit formation directly over it and drowning out the sound of the bomber.s

Blanketing smoke.—A chemical screen laid on, or in front of, the firing lines of the enemy. Blanketing smoke is used to cut down the enemy's vision and effect of his small-arms fire. It is also used as a screen between enemy and friendly forces to prevent enemy observation.

Blanket roll.—A folded or rolled blanket wrapped in the shelter half and usually containing poles and pines, undershirt, drawers, and socks. The blanket roll is usually carried in the pack.

Blast.—Sudden air pressure created by the discharge of a gun or the explosion of a charge.

Blast area.—A scorched area of ground in front of, and around, the muzzle of a gun, caused by repeated blasts.

Blast effect.—Destruction or damage to surface structures, etc., that is caused by the force of the explosion of a projectile or charge on it, or slightly above, the surface of the ground. Blast effect may be contrasted with the mining effect of a projectile or charge, which goes off beneath the surface.

Blasting cap.—A thin case inclosing a sensitive explosive such as mercury fulminate, used to set off another explosive charge. The explosive in the blasting cap is fired either by burning a fuze or by electricity.

Blast mark.—A worn area of the ground in front of a gun, caused by the force of the blast of firing. Unconcealed, it may give away the position of the gun.

Blister gas.—A poison gas that burns or blisters the skin or internal tissues of the body. Mustard as and lewisite are the common blister gases.

Blitzkrieg.—A rapid and well-timed offensive based on a coordinated advance of aircraft, armored units, and motorized troops, with the purpose of crushing all enemy resistance quickly.

Block.—*a.* A group of explosive units fastened together to go off at once.

b. An obstacle that prevents or hinders the advance of the enemy.

c. To hinder the movement of ground troops by placing obstacles across the route of advance.

d. To interfere with enemy radio broadcasts by transmitting on the same frequency.

Blockade.—*a.* The placement of armed forces so as to shut off an enemy from trade or communication with other countries.

   b. An armed force that maintains a blockade.

   c. To shut off enemy from trade or communication with other countries.

Blockhouse.—A fortified structure that has ports or loopholes through which gunfire is delivered.

Boat-carrying party.—*British term*: Used in connection with river crossings and only applied when folding boat equipment is used. It refers to the party that opens, carries, and launches the boats and includes the personnel required for the crew.

Bogie.—*a.* A railway undercarriage consisting of an axle and two wheels, or two axles and four wheels. Bogies are used to support the weight of a heavy body, such as an artillery gun or railway car.

   b. A roller or wheel that rides on the track of a tractor or tank, and takes up and distributes the weight of the vehicle along the track.

Bolt.—A sliding mechanism that closes the breech in some types of small arms. It usually contains the extractor and the firing pin.

Bolt mechanism.—A mechanical assembly in a bold action gun that includes the moving parts which insert, fire, and extract a round of ammunition.

Bomb.—*a.* A container filled with an explosive charge or chemical substance, usually dropped from aircraft. Some bombs contain both an explosive charge and a chemical substance. Bombs are exploded by contact or a time mechanism or a combination of the two. Common explosive bombs are demolition bombs to destroy heavy matériel or buildings, or fragmentation bombs to destroy personnel. Common chemical bombs include incendiaries and gas and smoke bombs.

   b. To drop one or more bombs from an aircraft on a target.

Bombardier.—A member of the crew of a bombardment airplane who operates the bomb sight and the bomb release mechanism.

Bombardment.—An attack made on a target by artillery fire or by bombs dropped from aircraft.

Bombardment airplane.—A combat airplane that is used for bombing missions; bomber.

Bombardment aviation.—That type of aviation whose primary mission is the attack of surface objectives; classified as *light, medium,* and *heavy. British equivalent*: **Bombers** or **bomber aircraft**.

Bomber Command.—*British term*: A subdivision of the Royal Air Force composed of all bomber squadrons, except those in the Coastal Command, in the United Kingdom.

Bomb line.—*British term*: A line usually delineated by well-defined geographical features, beyond which the air forces are free to attack any target. Targets on the friendly side of the bomb line will only be attacked by special arrangement with the ground forces. See **Bomb safety line**.

Bomb release line.—An imaginary line drawn around a defended area over which a bomber, traveling toward it at a constant speed and altitude, should release its first bomb to have it strike the nearest edge of the defended area; initial bomb release line.

Bomb safety line.—A line selected on the ground to insure reasonable safety to friendly troops from the effects of bombs dropped by supporting aviation. For their own security friendly ground troops do not pass this line until bombardment aviation lifts its fire. *British equivalent*: **Bomb line**.

Bomb sight.—An instrument in an airplane that is used to find the point in the flight of an airplane where dropping a bomb will cause the bomb to fall exactly on the target. A bomb sight calculates the speed and altitude of the airplane, the wind, and other variable factors, and allows for the effects of these to find the exact time to release the bomb so that it will hit the target.

Booster.—An additional explosive between the priming charge and the main charge to explode the main charge more quickly; booster charge. Sometimes a booster is part of the fuze.

Bore.—*a.* The inside of a gun barrel from the breechblock to the muzzle. Bore is used both for the inside surface of the barrel or tube of a gun, with its rifling, and also for the cylindrical space inclosed by the barrel.

b. The inside diameter of an engine cylinder.

Bore sight.—A device used to align the axis of the bore of a gun with an aiming point. A bore sight consists of a part attached to the muzzle of the gun and a part attached to the breech of the gun. The soldier using it sights through the bore sight to align the axis of the gun bore with an aiming point to which the gun sight is adjusted, so that the axis of the bore sight is aligned with the axis of the gun sight.

Boresight.—To sight thought the bore of a gun, with or without bore sights. Boresighting is done in lining up the axis of the bore of a gun with the sights.

Bound.—*a.* A single movement, usually from cover, made by troops, often under artillery fire or long-range small-arms fire. A forward move made by a series of bounds is called an **advance by bounds**. *British equivalent*: Same.

b. The distance marched by a unit when advancing in a successive series of moves. *British equivalent*: Same.

Boundary.—Lines designating the lateral limits of zones of action or of areas or sectors of defense. *British equivalent*: Same.

Bounding mine.—A type of antipersonnel mine, usually buried just below the surface of the ground. It has a small charge which throws the case up in the air; this explodes at a height of three or four feet, throwing shrapnel or fragments in all directions.

Box barrage.—A system of standing barrages inclosing an area. It is usually used to prevent the escape or reinforcement of enemy troops. *British equivalent*: Same.

Box magazine.—A box-like device that holds ammunition and feeds it into the receiver mechanism of certain types of automatic weapons.

Boys anti-tank rifle.—A British bolt-action magazine rifle, .55-caliber. It could penetrate 24-mm armor at normal angle of incidence at 100 yards, and 9-mm armor at a 40° angle at 500 yards.

Bracket.—*a.* The space between two shots or series of shots one of which is over the target and the other short of it, or one of which is to the right and the other to the left of the target.

b. To deliver fire that places a bracket on the target.

Bracket fire.—Fire delivered on a target in order to establish range limits over and short of the target or deflection limits to the right and left of the target. When the target has been inclosed by a bracket fire, the area containing the target can be covered by fire for effect, or the bracket can be narrowed by further adjustment fire until a bracketing salvo is laid on the target.

Branch.—*a.* A subdivision of any organization.

b. An arm or service of the Army. The Infantry, Cavalry, Medical Department, and Signal Corps are branches of the Army. *British equivalent*: **Arms of the Service.**

Brassard.—A band of cloth having vary ing colors or insignia on it, worn around the left sleeve as a sign that the wearer is assigned to certain special duties. Military police wear a brassard with MP in white letters.

Breach.—a. A gap or opening forcibly made in a fortification or position.

b. To create such a gap or opening.

Breach of arrest.—A military offense committed by an officer or enlisted man under arrest when, without permission, he leaves the limits of the area to which he has been restricted.

Break camp.—To pack all equipment and make ready to march, following a bivouac, a stay at a post, etc.

Breaking bulk.—*British term*: The process of subdividing bulk supplies received at a supply railhead on a formation section of the daily pack train to meet the requirements of each unit in a formation.

Break-through.—A penetration of the entire depth of a defensive system into unorganized areas in rear. *British equivalent*: Same.

Breastwork.—A field fortification that is dug deep enough, or put up high enough, to give protection to soldiers standing erect.

Breech.—The rear part of a bore of a gun, especially the opening that permits the projectile to be inserted at the rear of the bore.

Breechblock.—A movable steel block that closes the rear part of the barrel in a firearm.

Breech mechanism.—A mechanism that opens and closes the breech of a gun and fires the charge.

Bren gun carrier.—*British term*: see **Universal carrier.**

Bren light machine gun.—A .303-inch light machine gun that was the basic automatic weapon in the British Army. It is an air-cooled gas-operated magazine-fed machine gun; ordinarily fired from a bipod, but may be mounted on a tripod, an antiaircraft mount, or an armored automotive carrier.

Brevet commission.—An honorary commission in a higher grade given to officers as a reward for distinguished conduct. The honorary grade does not entitle the officer to higher pay or to greater command except when actually engaged in hostilities. *British*: Same.

Brevity code.—A relatively simple code used in radio communications, in which sentences and phrases are represented by code words or symbols. A brevity code provides some secrecy, but its main purpose is to save time in transmitting messages.

Bridge forming point.—*British term*: see ***Forming-up place.***

Bridgehead.—Position occupied by advance troops to protect the passage of a river or defile by the remainder of the command. *British equivalent*: Same.

Bridging party.—*British term*: Used in connection with river crossings and only when Kapok bridging equipment is used. It refers to the party that assembles, carries, launches, and maintains the bridge.

Brief.—A short, accurate summary of the details of a flight mission, given to the crew of a combat airplane just before it takes off to carry out its mission.

Brigade.—A tactical unit smaller than a division and larger than a regiment. Brigade is usually commanded by a brigadier general, and usually consists of troops of a single branch, such as artillery, infantry, or cavalry. In the British Army, a brigade takes on the tactical role of a ***regiment.***

Brigade reserve area.—*British term*: see ***Regimental area.***

Brigadier.—*British term*: A temporary rank above a colonel, held only while commanding a brigade or performing other duties for which the appropriate rank is a brigadier. Although a brigadier corresponds to the rank of brigadier general in the U.S. Army, it is improper to address such an officer as "General." The British rank of brigadier-general was abolished in 1920.

Brigadier general.—An officer in the Army who ranks next above a colonel and next below a major general. A brigadier general is the lowest grade of general officer. He is usually in command of a brigade in the ground forces, or a wing in the air forces.

Brigade reserve position.—*British term*: see ***Regimental reserve line.***

British Army.—*British term*: At the declaration of war on September 3, 1939, the British Army consisted of the Regular Army, the Territorial Army (corresponding to the National Guard in the United States), and several reserve forces. Soon thereafter all elements were consolidated into a single "British Army" and, except for certain legal differences, the distinctions between these several elements were eliminated.

Browning automatic rifle.—A self-loading rifle that can be adjusted to fire full automatic, semi-automatic, or hand-operated. A Browning automatic rifle is air-cooled, gas-operated, and magazine-fed. Late models had a bipod attached near the muzzle end to hold the gun steady when it was fired from a prone position on the ground.

Browning machine gun.—A machine gun firing .30-caliber or .50-caliber ammunition, used by the U.S. Army.

Buckling.—A breaking of a march column because the rear units crowd up on the front units.

Bugler.—A soldier who blows a bugle to give military signals. A bugler often acts also as a runner or messenger.

Bulk.—*British term*: Supplies received at the supply railhead on a formation section of the daily pack train are in bulk. These supplies are subdivided to meet the requirements of each unit in the formation; this subdividing is known as **breaking bulk**. The process of unloading, breaking bulk, and reloading on a unit basis is termed **refilling**. Bulk trains may also contain ammunition, engineer stores, or other commodities which are related to daily needs. See also **Bulk stock** and **Bulk supply**.

Bulk stock.—Full and unbroken packages of military supplies

Bulk supply.—Any kind of military supplies that is sent out in very large quantities. Sand, gravel, paint, gunpowder, etc., are examples of bulk supply. Bulk supplies are measured in terms of weight or volume rather than in terms of the number of units.

Bunker.—*a.* A mound of protective earth erected in front of a defended gun emplacement.

b. A storage space for fuel oil or coal on ships.

Bureau.—A subdivision of the War Department that carries out duties of a nonmilitary nature; for example, Bureau of Public Relations, National Guard Bureau. A bureau is distinguished from a division of the War Department, such as the Military Intelligence Division, that is primarily concerned with matters of a military nature.

Burst.—*a.* A series of shots fired by one pressure on the trigger of an automatic weapon.

b. An explosion of a projectile in the air or when it strikes the ground or target.

Burster.—An explosive charge used to break open and spread the contents of chemical projectiles, bombs, or mines.

Buzzerphone.—A portable field telephone and telegraph.

Cable block.—A road obstruction made by stretching a cable diagonally across a road so as to ditch a vehicle that hits the cable.

Cable ferry.—A set of cables strung over a stream or defile, over which equipment is moved from one bank to the other. The equipment is rigged on the cables and pulled across the stream or defile by a towline.

Cadence.—*a.* A uniform pace and time in marching.

b. The number of steps soldiers march per minute.

Cadet.—A student training in a military or aviation school for service as an officer.

Cadre.—(KAD ree). The key group of officers and enlisted men necessary to establish a unit.

Caisson.—A two-wheeled vehicle used for carrying artillery ammunition.

Caliber.—(*British*: Calibre) *a.* The diameter of the bore of a gun. A .45-caliber revolver has a barrel with an inside diameter of 45/100 of an inch.

b. The diameter of a projectile.

c. A unit of measure used to express the length of the bore of a gun or mortar. The number of calibers is found by dividing the length of the bore of the gun from the breechblock to the front of the muzzle by the diameter of its bore. A gun whose bore is 40 ft. long and 12 in. in diameter is said to be 40 calibers long.

Calibration.—*a.* Finding the correction elevation for a gun by firing the gun. Calibration is used to bring the fire of a gun into the same range as the other guns in the battery.

b. The measurement of wear in the bore of a gun in order to correct for a difference of muzzle velocity between it and the other guns of a battery.

c. Determining the corrections to be made in the readings of instruments used in precise measuring.

Calibration fire.—Experiment fire to determine the calibration corrections needed for the individual guns of the battery.

Call letter.—A code signal in communications used to identify a particular aircraft, radio station, or telegraph station.

Call sign.—A signal, usually a group of letters, or of letters and numerals, used for radio station identification. *British equivalent:* ***Code sign.***

Call (supply).—A demand for the delivery of supplies covered by credits.

Call to quarters.—A bugle call warning soldiers to go to quarters. Call to quarters is usually blown 15 minutes before taps is blown, and warns soldiers to to to quarters before beds are checked and lights are put out.

Call-up.—A set of signal used by a radio station to establish contact with another station.

Camouflage.—Work done for the purpose of deceiving the enemy as to the existence, nature, or location of material, troops, or military works. *British equivalent*: Same.

Camouflet.—(KAM a FLET *or* kam oo FLAY). *a.* A mine, bomb, or shell that explodes underground but whose explosion does not break the surface of the ground.

   b. A hole left behind the surface of the ground by such and explosion.

Camoufleur.—(kam a FLUR or kam oo FLUR). A man who conceals or disguises military objects by camouflage.

Camp.—Shelter consisting mainly of heavy tentage; a temporary location or station for troops; to put into camp; to establish a camp. *British equivalent:* Same.

Canalize.—To restrict an advance by natural or artificial obstacles and by fire into a narrow zone.

Candle.—A chemical container filled with a gas-producing or smoke-producing agent. The chemical filler is ignited, usually from a striker or spark at the top cover of the container, and produces a chemical warfare gas or smoke.

Cannibalization.—The using of equipment or parts from damage matériel to maintain other matériel

Cannon.—A fixed or mobile weapon on a mount that throws its projectile by the use of an explosive. Cannons are classified as guns, howitzers, and mortars.

Canteen.—*a.* A small container for water, coffee, etc., carried by a soldier.

   b. A former name for a shop or store for soldiers operated by the Army Exchange Service.

   c. A club or recreation center for soldiers, operated by a civilian organization such as the United Service Organizations.

Cantonment.—(kan TON munt). A group of temporary buildings especially erected for the shelter of troops. They are usually made of wood. A camp differs from a cantonment in that a *camp* usually has canvas tents for shelter. *British equivalent:* **Hutment.**

Capitulate.—To surrender on certain terms or conditions.

Captain.—*a.* An officer in the Army who ranks next above a first lieu-tenant and next below a major. A captain usually commands a com-pany, battery, troop, or flight.

 b. An officer in the Navy who ranks next above a commander and next below a commodore. A captain in the Navy has a rank equiva-lent of a colonel in the Army.

Capture.—To take or seize by force or stratagem.

Carbine.—A light rifle with a short barrel.

Carabineer.—A soldier armed with a carbine.

Cargador.—(KARG a dor). A person who supervises the loading and driving of pack animals.

Carriage.—*a.* A mobile or fixed support for a gun. It sometimes in-cludes the elevating and traversing mechanisms. In this meaning, usually called a gun carriage.

 b. The undercarriage of an aircraft.

Car post.—*British term: see* **Ambulance loading post.**

Carrier.—*a.* A motor vehicle for carrying men or matériel. The term is often combined with a word naming the special function for the carrier; for example, troop carrier or gun carrier. See also: **Loyd carrier** and **Universal carrier.**

 b. A harness or device for carrying small loads, such as chemical cylinders or gas masks.

 c. A person or thing that carries or spreads a disease. Carriers are often healthy persons who are immune to a disease, but carry its germs.

 d. A part of the mechanism of some automatic guns that helps to se the projectile in its proper firing position.

Casemate.—A bombproof structure, used as a gun emplacement, as a supply magazine, or for quartering troops.

Casualties.—Losses in numerical strength by death, wounds, sick-ness, discharge, capture, or desertion. *British equivalent:* Same.

Casualty agent (chemical).—*A* material of such physical and chemi-cal characteristics that a dangerous or killing concentration can be set up under conditions encountered in the field. *British equivalent:* **Poisonous gas.**

Casualty clearing station.—*British term:* A location that receives sick and wounded from field ambulances, and direct from troops in the vicinity.

Casualty collecting point.—*British term*: A medical post formed by a section, or part of a section, of a field ambulance, to which casualties are brought from the regimental aid post, and to which ambulance cars of the field ambulance are sent to collect casualties.

Cavalry.—Highly mobile ground units, horse, motorized, or mechanized.

Cave shelter.—An underground shelter for troops, dug out from a bank or hillside, with undisturbed ground over it. A cave shelter differs from a **cut-and-cover shelter**, which is an open pit with an artificial cover.

Ceiling.—*a.* The distance between the lower level of a cloud bank and the ground.

    b. The greatest height to which an aircraft can go.

Censorship.—Measures taken to prevent the leakage of information; they are applied to private communications, photography, press dispatches, radio broadcasts, publications, and all communications.

Center.—The middle point or element of a command. If the number of elements considered is even, the right center element is considered the center element. *British equivalent*: None.

Center of resistance.—A point in the main defensive system at which troops are concentrated to repel enemy attacks. A center of resistance supports, and is in turn supported by, other centers of resistance.

Centre line.—*British term*: The route along which the headquarters of a formation, unit, or sub-unit moves. See also **Axis of signal communication**.

Chain of command.—A series of commanding officers through whose ands must pass orders, especially field or combat orders, and instructions which travel from a commander to a subordinate. Also called a **command channel**.

Challenge.—*a.* A command of a sentry to anyone approaching a sentry's post to halt and identify himself.

    b. To order anyone approaching a sentry's post to halt and identify himself.

Chamber.—The part of a gun in which the charge is placed. In a revolver it is a hole in the cylinder, in a cannon it is part of the breech, and in most rifles and automatic weapons it is the rearmost part of the bore.

Channel.—A route of official communication between headquarters or commanders of military units. Also called **Military channel.**

Charge.—*a.* An explosive used in firing a gun. A propelling charge throws a projectile from a gun. A bursting charge breaks the casing of a projectile to produce demolition, fragmentation, or chemical action.

> *b.* A violent final rush in an attack or assault.

> *c.* To make such a rush.

Check concentrations.—Registration of fire on easily identified points throughout the zone of fire, from which transfers can be made to targets of opportunity. See also **Fire for adjustment** *and* **Registration.** *British equivalent*: **Registration.**

Chemical agent.—A substance useful in war which, after release and acting directly through its chemical properties, is capable of producing a toxic effect, a powerful irritant effect, a screening smoke, or an incendiary action.

Chemical cylinder.—A cylindrical tank from which chemical agents are released through a valve by means of internal gas pressure. *British equivalent*: Same.

Chemical land mine.—A container of persistent gas employed with a detonator to contaminate surrounding ground and vegetation. *British equivalent*: Same.

Chemical officer.—An officer, usually a member of the Chemical Warfare Service, who serves on the staff of a commander of a division or higher uni and advises the commander and staff on all matters regarding chemical warfare. In units small than a division, the staff officer having these duties is call the **gas officer.**

Chemical warfare.—Tactics and technique of conducting warfare by use of chemical agents. Attacking agents are gas, fire, and smoke, and they are spread by means of shells, bombs, grenades, cylinders, smoke generators, and flame throwers. Defenses are neutralizing or decontaminating agents and devices, such as gas masks and special clothing. *British equivalent:* Same.

Chevron.—A cloth design shaped like an inverted-**V** or **V**, worn on the sleeve to indicate the rank, wounds, or length of service in a combat area.

> **Grade chevron.**—A khaki or olive-drab close shaped like an inverted-**V**, to indicate the rank of an enlisted man above the grade of private. A grade chevron is worn with the point up, on the upper part of both sleeves.

War service chevron.—A gold-colored cloth design shaped like a **V**, to indicate six months' foreign service in World War I. It is worn with its point down, on the lower part of the left sleeve.

Wound chevron.—A gold-colored cloth design shaped like a **V**, to indicate wounds received in World War I. A wound chevron is worn with its point down, on the lower part of the right sleeve.

Chief of service.—The senior general officer in one of the supply or administrative services of the Army, such as the Finance Department or the Chemical Warfare Service. The chief of service is charged with preparing plans and policies for the organization, training, equipment, and operation of the service. There were no longer any chiefs of arms, such as the Cavalry or Field Artillery.

Chief of Staff.—The senior general staff officer, detailed as such, on duty with the staff of a division or higher unit. See **General Staff.** *British equivalent:* No exact equivalent.

Cipher.—A method of secret writing that substitutes other characters for the letters intended or transposes the plain text letters or employs both these processes. *British equivalent:* Same

Circulation map.—A map showing the measures for traffic regulation. *British equivalent:* **Traffic map.**

Citation.—*a.* A specific mention in orders or dispatches.

b. A public commendation or decoration for unusual achievement or gallant action.

c. A reference to legal or other authorities in support of rules and regulations prescribed in Army manuals.

Class I supplies.—A class of supplies consisting of those articles which are consumed at an approximately uniform daily rate irrespective of combat operations or terrain, and which do not necessitate special adaptation to meet individual requirements; such as rations and forage.

Class II supplies.—A class of supplies consisting of those authorized articles for which allowances are established by Tables of Basic Allowances; such as clothing, gas masks, arms, trucks, radio sets, tools, and instruments.

Class III supplies.—A class of supplies consisting of engine fuels and lubricants, including gasoline for all vehicles and aircraft, Diesel oil, fuel oil, and coal.

Class IV supplies.—A class of supplies consisting of those articles which are not covered in Tables of Basic Allowances and the demands for which are directly related to the operations contemplated or in progress (except for articles in classes III and V); such as fortification materials, construction materials, and machinery.

Class V supplies.—A class of supplies consisting of ammunition, pyrotechnics, antitank mines, and chemicals.

Clear (verb).—To pass a designated point or line. Refers to the tail of a unit. *British equivalent*: Same.

Clear (in the).—The sending of messages, orders, or instructions in plain (uncoded) language. *British equivalent*: Same.

Clearing station.—The corps or division medical installation where sick and wounded are assembled from the collecting stations and aid stations, sorted, treated if necessary, and turned over to the army for further evacuation. (Formerly called **hospital station**.) *British equivalent*: **Casualty clearing station**.

Clock method.—A method of calling shots by reference to the figures on an imaginary clock dial assumed to have the target at its center; clock-faced method; clock system. Thus, a shot directly above the target is at 12 o'clock. The same method was sometimes used to name the direction of the wind; for example, a wind directly from the left is a 9 o'clock wind.

Close envelopment.—An enveloping maneuver intended to strike the flan of the enemy and surround the enemy forces on the flank. It differs from a **wide envelopment**, which is an enveloping maneuver that starts from the enemy position and usually is directed at an objective far in the rear of the enemy front lines.

Close order.—Any formation in which units are arranged in line or column with normal or close intervals and distances.

Close reconnaissance.—A reconnaissance of a region near at hand. Exploration of objectives that lie outside the immediate striking range of a force is called **distant reconnaissance**.

Close support.—The effective air or ground support given at close range by one combat arm or unit to another combat arm or unit.

Coastal Command.—*British term*: A subdivision of the Royal Air Force based in the United Kingdom specifically charged with general reconnaissance over the sea by shore-based aircraft. It also included a relatively small striking force of shore-based bomber and torpedo-bomber aircraft, and an appropriate number of long-range fighter aircraft to provide cover against attacks on shipping in focal areas outside of the range of short-range Fighter Command aircraft.

Coastal force.—A naval force which may be organized to operate within the coastal zone to meet a special situation in which naval local defense forces are inadequate to carry out the Navy's functions in coastal waters. *British equivalent*: Same.

Coastal frontier.—A geographic division of friendly coastal area established for organization and command purposes in order to insure effective coordination between Army and Navy forces engaged in coastal defense. *British equivalent*: None.

Coastal frontier defense.—The organization of the Army and Navy forces and their installations assigned to the defense of coastal frontiers. *British equivalent*: **Coast defence force.**

Coastal zone.—The whole area of the navigable waters adjacent to the seacoast. It extends seaward to cover coastwise sea lanes and focal points of shipping approaching the coast. *British equivalent*: Same.

Coast defence artillery.—*British term*: see **Seacoast artillery (Coast Artillery Corps)**.

Coast defence force.—*British term*: see **Coastal frontier defense**.

Coast route.—*British term*: see **Coastwise sea lane.**

Coastwise sea lane.—The water area adjacent to the seacoast. It includes all the usually traveled routes of coastwise shipping. *British equivalent:* **Coast route.**

Cobelligerent.—One of a number of nations or political units engaged in war against a common enemy.

Cocking lever.—A device in a gun's action that prepares the gun to be fired by moving the firing pin back into firing position.

Code.—A method of secret writing that substitutes arbitrary groups of symbols given in a code book as equivalents of whole sentences, phrases, words, letters, or numbers. *British equivalent*: Same.

Code group.—Two or more letters or numbers that have special meaning in a code system.

Code sign.—*British term*: A secret group of letters, or letters and figures, which is changed at frequent intervals, used to identify a headquarters. See **Call sign.**

Code word.—*British term*: A pre-arranged secret word used to convey instructions or information.

Collecting point.—A point designated for the collection of prisoners of war or stragglers. *British equivalent:* **Collecting post** (for prisoners of war only).

Collecting post.—A medical station in the forward combat zone where battle casualties are prepared to be sent to collecting stations in the rear. Collecting posts differ from collecting stations, in that the station is a more elaborate installation.

Collecting station.—*a.* An establishment located in the forward combat zone for the purpose of collecting and receiving casualties from aid stations and units, and preparing them for further evacuation by ambulance.

   b. Any place in the forward area for collecting and sorting salvaged materials.

Colonel.—An officer in the Army who ranks next above a lieutenant colonel and next below a brigadier general. A colonel holds the highest rank of field officer, and usually commands a regiment.

Color.—A flag, especially one carried by dismounted units; colors. ***To the color*** is a bugle call sounded as a salute to the flag, or to the President, the Vice-President, and ex-President, or a foreign chief magistrate. Also called ***to the colors*** or ***to the standard.***

Column.—A formation in which the elements are placed one behind another. A ***march column*** comprises all elements of a command marching on one route under the control of one commander, including such forward, flank, and rear security forces as may be employed. *British equivalent*: Same or ***line ahead.***

Combat arm.—A branch of service used in actual fighting, such as the Infantry, Coast Artillery Corps, or Army Air Forces.

Combatant.—A soldier or unit assigned to duty as an active fighter, as distinguished from one on duty in any of the services, such as administration, supply, or medical care.

Combat aviation.—A term applied to bombardment and pursuit aviation.

Combat car.—A light armed and armored track-laying vehicle designed for active fighting. A light tank is one kind of combat car.

Combat element.—Troops that actually take part in fighting, distinguished from troops engaged in supply or administration.

Combat echelon.—The principal element of offensive or defensive power. For example, the infantry echelon in defense occupying the principal battle position. *British equivalent*: ***Fighting group.***

Combat element.—Troops that actually take part in fighting, distinguished from troops engaged in supply or administration.

Combat engineers.—Troops of the Engineer Corps organized, trained, and equipped for actual battle service in addition to construction and repair work.

Combat intelligence.—Military intelligence produced in the field, after the outbreak of hostilities, by the military intelligence section of GHQ and military intelligence sections of all subordinate units.

Combat liaison.—A system of maintaining contact and communication between units during fighting, in order to secure proper coordination.

Combat orders.—Oral, dictated, or written orders issued by a superior to a subordinate unit, covering any phase of operations in the field. Combat orders include field orders. letters of instruction, and administrative orders. *British equivalent*: **Operation orders** or **operation instructions**, or (for division and higher levels) **administration orders (Adm Orders)**.

Combat outpost.—The outpost or security detachments established by subordinate commanders (company or battalion) when the distance of the security echelon from the main line of resistance is so reduced that the security troops can be more effectively coordinated with, and supported by, the combat echelon than a separate outpost under the control of higher commanders. For example, for a company it usually consists of one or more small outguards posted within a close range of the position; for a battalion, several platoons sent forward as outguards beyond close range and within the effective range of battalion heavy weapons. *British equivalent*: **Outpost**.

Combat patrol.—A tactical unit sent out from the main body to engage in independent fighting; detachment assigned to protect the front, flank, or rear of the main body, by fighting if necessary.

Combat reconnaissance.—The reconnaissance of the enemy in immediate contact with one's own forces, preliminary to, or during combat.

Combat team.—A nonorganic grouping of two or more units of different arms, such as an infantry regiment, a field artillery battalion, and a combat engineer company. *British equivalent*: **Group** (with the basic organization designated before it: e.g., brigade group).

Combat unit.—A unit trained and equipped for fighting as an independent group.

Combat unit loading.—*See* **Unit loading**.

Combat vehicle.—A self-propelled, armed vehicle, with or without armor, manned by combat personnel.

Combat zone.—The region where fighting is going on; the forward area of the theater of operations where combat troops are actively engaged. It is divided into army, corps, and division areas and extends from the front line to the front of the ***communications zone.*** *British equivalent:* ***Forward area.***

Combined arms.—More than one tactical branch of the Army used together in operations.

Combined operations.—The tactics of the combined or associated arms, such as the Infantry, Cavalry, Field Artillery, Corps of Engineers, Air Corps, or any two or more of them. Joint operations, as by two or more allies, by the Army and Navy, etc. *British equivalent:* Same (but applied only to operations carried out by sea and land forces or by sea, land, and air forces.

Command.—*a.* The authority which an individual exercises over his subordinates by virtue of rank and assignment. *British equivalent:* Same.

b. The direction of a commander expressed orally and in the prescribed phraseology. *British equivalent:* Same.

c. A body of troops or a locality under the command of one individual. One of the essential elements of military organization, the other two being combat and supply elements. *British equivalent:* Same.

d. The vertical height of the fire crest of the parapet above the original natural surface of the ground. *British equivalent:* Same.

e. The vertical height of any ground over other ground in its vicinity. To order or exercise command. *British equivalent:* Same.

f. A tactical and administrative unit of the U.S. Army Air Forces, which may include wings and groups or a number of groups alone. It is designated according to its primary function as a fighter, bomber, air support, troop carrier, transport, or air service command.

g. One of the large administrative subdivisions of the Army Air Forces, such as the Materiel Command, Air Transport Command, Training Command.

Command car.—A motor vehicle, usually armed and armored, equipped with facilities to assist in the exercise of command therefrom. *British equivalent:* ***Armoured command vehicle.***

Command channel.—See ***Chain of command.***

Command echelon.—See ***Command element.***

Commandeer.—To take over private property or the services of a private individual for military or public use.

Command element.—A group of officers and enlisted men who from the directing and coordinating head of a tactical or administrative unit, usually including the unit commander, his staff, and attached personnel, that performs duties in relation to administration, intelligence, communications, and other necessary activities; command echelon. Also called **headquarters**.

Commander.—*a.* An officer in command of a post or unit, such as the commanding officer of a regiment. In this meaning, also called commanding officer.

b. An officer in the Navy who ranks next above a lieutenant commander and next below a captain. A commander in the Navy is equivalent in rank to a lieutenant colonel in the Army.

Commander in Chief.—A person who has complete command of the armed forces of a country. The President of the United States is Commander in Chief of all the armed forces of the United States. Sometimes a high ranking officer of the Army or Navy is called commander in chief of a theater or war, a fleet, etc.

Commander's group.—A subdivision of the forward echelon containing certain key officers and men who usually accompany the commander. During combat it. may be combined with the command post.

Commanding general.—A title for any officer of brigadier general's rank or higher who is in command of a service, post, school, tactical unit, theater of operations, port of embarkation, hospital unit, etc.

Commanding ground.—High land which overlooks the surrounding region. Because of its superior location for observation and fire it controls the lower ground. *British equivalent*: Same.

Commanding officer.—An officer in command of a post or a unit, such as the commanding officer of a regiment. Also called **commander**.

Command net.—A radio, telegraph, or telephone system of communications over which commands are sent to various headquarters and units.

Commando.—A soldier specially trained to make surprise attacks on enemy territory. Commandos act in small groups, making rapid attacks and withdrawing. The name **Commando** is used by the British; the corresponding American term for soldiers of this kind is **Ranger**.

Command post.—*a.* The staff agencies and command facilities immediately required by the commander for assistance in tactical operations; the forward echelon of a unit's headquarters. The locations of such agencies and facilities. *British equivalent:* **Advanced (or forward) HQ.**

    b. British term: see **Advance command post.**

Command post exercise.—A tactical field exercise in setting up a command post, carried on by a headquarters staff and communication personnel. Command post exercises vary in nature from an exercise that closely resembles a map maneuver to exercises in which all command posts and communications are actually installed on the ground.

Commendation.—A written citation signed by a commander, in recognition of praiseworthy action not meriting a decoration.

Commercial loading.—A method of loading as ship or an aircraft so as to make the best use of all available space, without attempting to keep units and their equipment together. It is used when, and if, there is time to issue equipment to troops after they have landed.

Commissary.—An office where subsistence stores are issued or sold. See **Post exchange (PX).** *British equivalent:* **Grocery shop and store**, **ration shop**, and **local produce store** (part of an institute, usually run by the NAAFI).

Commission.—*a.* A written order giving a person rank and authority as an officer in the Army or Navy.

    b. The rank and authority given by such and order.

    c. To put in service or use; to make ready for service or use; for example, to commission an aircraft or ship.

Commissioned officer.—An officer in any of the armed services who holds a commission. In the Army, a person who has been appointed to the rank of second lieutenant or higher is a commissioned officer.

Commissioned warrant officer.—An officer in the Navy who ranks next above a warrant officer, junior grade, and next below an ensign.

Commit.—To make a positive decision to send a particular unit into an engagement or attack; to send a unit into battle.

Commodore.—An officer in the Navy who ranks next below a rear admiral and next above a captain. A commodore is equivalent in rank to a brigadier general in the Army.

Communications.—*a.* The means, methods, and routes of sending messages, such as telephone and telegraph, radio, or signal systems. *British equivalent:* Same.

   b. The routes and transportation for moving troops and supplies, especially in a theater of operations. *British equivalent:* Same.

Communications zone.—That part of the theater of operations between its rear boundary and the rear boundary of the combat zone, containing the lines of communication, the establishments of supply and evacuation, and other agencies required for the immediate support and maintenance of the field forces in the theater of operations. *British equivalent:* **Line of communications area.**

Communication trench.—A trench designed primarily to provide cover for personnel moving from one part of an entrenched position to another. Also called a **connecting trench** or a **communicating trench.** See also **Approach trench.** *British equivalent:* Same.

Communique.—(kum YOON a KAY). Information issued officially, especially an account of recent activities or events.

Company.—The basic administrative and tactical unit of most branches of the military service, larger than a platoon, smaller than a battalion. A company is equivalent to a battery of artillery, a troop of cavalry, or an aviation squadron, and is usually commanded by a captain.

Company aid men.—Soldiers of the Medical Corps assigned to a company for the purpose of providing emergency treatment and caring for casualties in combat. *British equivalent:* **Stretcher-bearers.**

Company clerk.—A person, usually a noncommissioned officer, assigned to help the first sergeant in doing the clerical work of a company.

Company defense area.—*See* Defense area. For example, a company defense area is a section of the battalion defense area assigned to a rifle company by the battalion as its task in the all-around defense of the battalion area.

Company discipline.—See **Company punishment.**

Company grade.—A classification of those officers normally serving in a company. It applies to lieutenants and captains.

Company officer.—Any officer ranking below a major and serving in a company. Captains, first lieutenants, and second lieutenants are company officers. See also: **Subaltern.**

Company punishment.—Light punishment and other corrective measures imposed by a company commander without resort to a court-martial; company discipline.

Compartments of terrain.—An area of terrain inclosed on at least two opposite sides by terrain features such as ridges, woods, cities, or bodies of water, which prevent ground observation and direct fire into the area. Also called **Terrain compartment**. *British equivalent*: None.

Complement.—*a*. The full, authorized strength of a military unit or post, including officers, men, and matériel.

 b. Extra units of various types attached to a given unit for adding to its services or operations.

Component.—One of the parts of the Army of the United States, such as the Regular Army or National Guard.

Composite photograph.—The picture that results from the joining together of the vertical and transformed oblique photographs made by a multilens camera. *British equivalent*: Same.

Concealment.—The state or condition of being hidden from the enemy's view. Any object affording protection from the view of the enemy. Concealment offers protection from observation only; **cover** offers protection from gunfire. *British equivalent*: Same.

Concentrated fire.—Fire from a number of guns, directed at a single point or small area; fixed fire; point fire. Concentrated fire is heavy fire directed at a strong point, fixed fortification, or enemy concentration; **distributed fire** covers a wider area more lightly.

Concentration.—*a.* An assembly of troops in a particular locality, on mobilization for training, attack, or defense. *British equivalent*: Same.

 b. The amount of toxic chemical vapor in a given volume of air at any particular time and place. *British equivalent*: Same.

 c. A volume of fire placed on an area within a limited time. *British equivalent*: **Massed fire**.

 d. British term: see **Assembly**.

Concentration area.—*a.* An area, usually in the theater of operations, where troops are assembled before beginning active operations.

 b. A limited area on which a volume of gunfire is placed within a limited time.

Concentration march.—A march made by troops to reach an assembly point or area at which they will join other troops.

Concertina.—A cylinder-shaped, portable wire entanglement, which can be folded or pushed together for carrying, and extended for use. Concertinas are normally used to increase the strength and effectiveness of other types of entanglements.

Conduct of fire.—All operations connected with the preparation and actual application of effective fire upon a target. Also called ***fire control.***

Conference call.—A telephone call in which one individual desires to obtain telephone connection with two or more other individuals at the *same* time in order to transmit instructions or information to all parties simultaneously. *British equivalent:* ***Multiple call.***

Confidential.—A classification of information or material which is allow to reach only those persons who need it to carry out their duties. Information not public during World War II was classified as restricted, confidential, or secret. Confidential information, although not secret, was less freely circulated than restricted information.

Confinement.—Being kept in prison or under guard in a specific place. Confinement does not include restriction to barracks or other area not under guard.

Connecting group (or file).—Any group (or file) used to maintain contact between separated forces or elements. *British equivalent:* ***Connecting file.***

Connecting trench.—A trench by which men can move from one fire trench to another, or from one part of a trench system to another; communicating trench; communication trench. See also ***Approach trench.***

Conscientious objector.—A person whose beliefs forbid him to take an active part in warfare. A conscientious objector, if taken into the Army, is assigned to noncombatant duties.

Conscript.—A man drafted for military service.

Conscription.—Drafting for military service.

Consolidated return.—A report made by combining the reports from several subordinate organizations or stations.

Consolidation of position.—The act of organizing and strengthening a position recently captured. *British equivalent:* Same.

Constitute.—To establish a new unit on the active or inactive list of the Army. The new unit is set up on the list but it has no physical existence until it has personnel and equipment assigned to it. Constitute differs from **activate**, which means to establish a new unit on the active list of the Army and assign to it personnel and equipment.

Contact (or encounter) battle.—*British term*: see **Meeting engagement.**

Contain.—To stop, hold, or surround the forces of an enemy; to cause the enemy to center his activity on a given front and prevent his withdrawing any part of his forces for use elsewhere. *British equivalent*: Same.

Containing action.—An attack designed to hold the enemy to his position or to prevent him from withdrawing any part or all of his forces for use elsewhere. See also **Holding action.** *British equivalent:* Same.

Containing force.—A body of troops whose mission is to hold an enemy force in check or position. *British equivalent:* Same.

Contaminated area.—An area contaminated with persistent gas.

Contaminated vehicle area.—An area to which vehicles are sent for decontamination.

Contingent barrage.—A barrier of gunfire planned a ready to be employed in a possible emergency.

Contingent zone.—An area within the field of fire, other than the normal zone, within which a unit may be called upon to fire. *British equivalent*: None.

Continuing attack.—See **Attack, continuing**

Continuous fire.—Fire at the normal rate without repeated orders to fire and without interruption for corrections of adjustment or for other reasons, until an order to cease fire is received.

Contour.—*a.* An imaginary line on the ground joining all points of the same elevation.

  b. A line on a map representing this elevation.

Contour interval.—The difference in elevation of two or more adjacent contours; vertical interval. *British equivalent:* **Vertical interval.**

Contour map.—A map showing heights above sea level by lines that connect points on a land surface having the same height.

Contraband of war.—Goods, supplied by neutral nations to any country at war with another, that either warring country has a right to seize. Ammunition is always a contraband of war.

Controlled mosaic.—An assembly of two or more overlapping vertical photographs oriented with respect to each other and to a framework of points appearing on the photographs whose locations on the ground have been definitely determined. *British equivalent:* Same.

Control officer.—An officer moving at the head of a march unit to set the pace and to insure keeping to the assigned route and march schedule.

Control of the air.—Total neutralization of enemy's aviation. Freedom of all friendly air and surface forces to operate without hostile air opposition.

Control point.—An agency established by a unit at a convenient point on the route of its trains at which information and instructions are given and received in order to facilitate and regulate supply or traffic. *British equivalent:* **Meeting point** (for supply) or **traffic control post.**

Convalescent camp.—A branch of a hospital center where sick or injured soldiers complete their recovery after leaving the hospital.

Convalescent hospital.—A fixed or mobile hospital for patients who no longer need active treatment.

Converging attack.—An attack from different directions delivered upon one point or place.

Converging fire.—Fire from a number of different guns directed at the same spot.

Convoy.—Any group of transportation temporarily organized to operate as a unit during movement. To escort. To accompany for the purpose of protecting. *British equivalent:* Same.

Convoy unit loading.—*See* **Unit loading.**

Cooperation.—The act of working together for the accomplishment of a common end. *British equivalent:* **Co-operation.**

Coordinated attack.—An attack in which each element has its own mission, planned so that the activities of all elements make an effective whole, rather than employing elements as they become available, as in a **piecemeal attack.**

Coordination.—The act of supervising, regulating, and combining. to gain the best results. *British equivalent:* Same.

Cordite.—A slow-burning powder made of guncotton, nitroglycerin, and mineral jelly, used in artillery projectiles.

Corduroy road.—A road surfacing for soft or marshy ground, made by laying whole or split logs crosswise, either on the natural surface or on logs or planks placed lengthwise.

Corporal.—*a.* A noncommissioned officer of the fifth rank in the Army who ranks next above a private, first class and next below a sergeant. He is usually in command of a squad.

    b. A title of address for a technician, fifth grade.

Corporal of the guard.—A noncommissioned officer of the guard who instructs and posts sentries, changes reliefs, and is change of one of the reliefs.

Corps.—*a.* A tactical unit larger than a division and smaller than an army. A corps usually consists of two or more divisions together with auxiliary arms and services. In this meaning, in the U.S. Army it was previously call an *army corps*.

    b. One of the branches of the Army; such as Coast Artillery Corps, Corps of Engineers, Quartermaster Corps, and Medical Corps of the Medical Department.

Corps area.—See *Service command*.

Corps artillery.—Artillery that is assigned to a corps, but not as a unit within it.

Corps of Engineers.—A branch of the Army that constructs and operates shelters, fortifications, bridges, and other structures, and many kinds of mechanical equipment.

Corps of Military Police.—A branch of the Army that functions as a police force to protect persons and property and to maintain order in areas under military control. The Corps of Military Police is under the supervision of the Provost Marshal General.

Corps troops.—Troops assigned or attached to a corps, but not a part of one of the divisions that make up the corps. They are assigned or attached for special purposes, and usually for a limited time.

Corresponding range.—The shortest range at which fire can safely be delivered over the heads of troops. It is the range from a gun to a point where the curved path of a projectile, passing safely over the heads of troops, will strike on level or uniformly sloping ground.

Corridor.—A compartment of terrain of which the longer dimension lies generally in the direction of movement or leads toward an objective. For example, an avenue of approach having natural terrain features on its two flanks which limit observation and direct fire from positions outside the corridor constitutes favorable lines of advance for friendly or hostile forces. *British equivalent:* Same.

Cossack post.—An outguard consisting of four men posted as an observation group, with a single sentinel in observation, the remaining men resting nearby and furnishing the reliefs for the sentinel. *British equivalent:* **Observation post, listening post** by night (no definite number of soldiers).

Counterattack.—An attack by part or all of the defending force against a hostile attacking force for the purpose of regaining ground lost or for cutting off or destroying hostile elements. It is mostly a defensive action with only temporary and local offensive action to seize limited objectives. A counterattack differs from a counteroffensive, which is an aggressive action on a large scale, undertaken by a defending force to seize the initiative from the attacking force. *British equivalent:* **Immediate counter-attack** (launched before consolidation of position) or **deliberate counter-attack** (launched after consolidation of position).

Counterbarrage.—A barrage laid down in opposition to an enemy barrage.

Counterbattery fire.—Artillery fire delivered for the neutralization or destruction of enemy batteries in position. *British equivalent:* **Counter-battery fire.**

Counterespionage.—Measure taken to prevent espionage by the enemy.

Counterintelligence.—Measures taken to destroy the effectiveness of the enemy's intelligence system.

Countermand.—*a.* To withdraw or cancel an order or command.
   b. To call back or stop by contrary order; order back.

Counteroffensive.—An aggressive action on a large scale undertaken by a defending force to seize the initiative from the attacking forces. The purpose of a counteroffensive is to stop the offensive of the enemy and destroy his forces. A counteroffensive differs from a **counterattack**, which is mostly a defensive action with only temporary and local offensive action to seize limited objectives. A counteroffensive usually follows a counterattack. *British equivalent:* **Counter-offensive.**

Counterpreparation.—Prearranged fire delivered in a defensive action just prior to the enemy attack for the purpose of breaking up the attack or reducing its effectiveness. *British equivalent:* **Defensive fire** (term "counter-preparation" abolished).

Counterreconnaissance.—All measures taken to screen a command from any form of enemy observation or to neutralize its effectiveness. *British equivalent*: **Counter-reconnaissance.**

Counterreconnaissance screen.—Units of an attacking or maneuvering force that protect the main force from effective enemy reconnaissance. Cavalry and motorized units are often used as a counter-reconnaissnce screen.

Countersign.—A password given in answer to the challenge of a sentinel. The countersign includes the challenge, when secret, and the sentinel's reply to the password.

Course of action (or courses open).—*British term*: see **Lines of action.**

Court-martial.—*a.* A military court; a court of army officers that tries military personnel accused of offenses against military law and regulations. The number of officers sitting in judgment varies with the three kinds of court-martial: general, special, and summary. *British equivalent*: Same (but classified the courts-martial as: general, district, and field general).

b. Trial by such a court.

c. To try by such a court.

Court of inquiry.—A board of officers appointed by a competent authority to examine any transaction, action, accusation, or charge against a member of a command.

Cover.—*a.* Natural or artificial shelter or protection from fire or observation, or any object affording such protection. **Cover** offers protection from enemy fire; **concealment** offers protection from observation only. *British equivalent*: Same.

b. The vertical relief of a trench measured from the bottom, or from the trench board, to the top of the parapet. *British equivalent*: Same.

c. To protect or provide security for another force or a locality. *British equivalent*: Same.

Coverage.—The area covered in any one exposure by an aerial (or air) photograph. *British equivalent*: None.

Covered approach.—*a.* Any route that offers protection against enemy observation or fire.

b. An approach made under protection furnished by other forces or by natural cover.

Covered approach march.—An approach march which is protected by forces sufficiently strong to provide security against hostile ground attack.

Covering fire.—*a.* Fire used to protect troops when they are within range of the enemy's small arms. Covering fire is usually artillery fire protecting infantry.

b. British: Fire by one unit or arm to engage the enemy's attention and force him to seek cover in order that another unit or arm may advance or retire.

Covering force (or detachment).—*a.* Any body or detachment of troops which provides security for a larger force by observation, reconnaissance, attack, or defense, or any combination of these methods. *British equivalent:* Same, also see **March outpost** and **Security detachment.**

b. British: In combined operations, the force which lands first to secure the **covering position.**

Covering position.—*British term:* In combined operations, a position to be occupied by the covering force, at such a distance from the landing places as will reasonably ensure immunity from observed artillery fire upon landing places and transports.

Covering troops.—*British term:* A major detachment of troops operating in advance of a main force to prevent or delay interference with the activities of the main force, especially during the preparation of a defensive position or during a strategical withdrawal.

Cover position.—A position immediately in rear of the fire position which affords protection to the riflemen, or to a weapon, from hostile flat-trajectory fire.

C ration.—A field ration that is a cooked balanced ration in cans. Each ration consists of three cans of prepared meats and vegetables and three cans of crackers, and soluble coffee. As this ration is not perishable, it is suitable for use as a unit reserve or as an individual reserve.

Crawl trench.—*British term:* A shallow trench, 3 ft. 6 in. wide, semicircular in section and 18 in. deep at the center. It is used as a first stage to connect weapon pits and posts across the front, and facilitates the rapid development of an extensive trench system.

Credit.—An allocation of a definite quantity of supplies which is placed at; the disposal of the commander of an organization for a prescribed period of time. *British equivalent:* **Reserve supplies.**

Crest.—*a.* The summit or highest line of a ridge. The actual or topographical crest. *British equivalent:* Same.

b. British usage also for **Military crest.**

Critical points.—Selected terrain features along a route of march with respect to which instructions are given to serials for the purpose of controlling the movement. *British equivalent:* **Locations for sector controls** or **traffic control points.**

Cross compartment.—A narrow section of terrain whose length lies across the direction of movement of a force, or is parallel to the front. Also called **cross corridor.**

Cruiser tank.—*British term:* The armored vehicle of the British armoured brigades, they are relatively fast and maneuverable, and sacrifice armor to speed and armament. British cruiser tanks vary considerably in weight, certain models weighing the same as U.S. light tanks, other weighing the same as U.S. medium tanks. The normal role, however, of British cruiser tanks is similar to that assigned to U.S. medium tanks.

Cruising.—Operation of tanks on the objective to keep down hostile fire until arrival of attacking foot or other troops.

Cryptanalysis.—Analysis of secret messages; the process of reading a ciphered or coded message without having the key.

Cryptogram.—A message written in secret code or cipher, or in combination of code and cipher.

Crypotgraph.—To translated a message from plain text to secret code or cipher. Also called encrypt.

Cryptographic security.—That form of signal-communication security which deals with the provision of technically sound cryptographic systems, their proper use, and their careful protection. *British equivalent:* Same.

Cryptography.—The science which embraces the methods and devices used to convert a written message into code or cipher. *British equivalent:* Same.

Cupola.—An armored turret, especially the small hatch opening into the top of the main turret of certain types of tanks.

Curtain of fire.—A screen or barrier of fire that is laid down along a line or in an area. A curtain of fire is usually laid down to protect attacking forces from enemy attack or observation.

Customs of the service.—Military practices not covered by regulations but regarded as binding because of tradition.

Cut-and-cover shelter.—A shelter constructed by digging an open pit, then providing an overhead cover. A cut-and-cut shelter is different from a **cave shelter**, which is completely underground, or a *surface shelter*, which is completely above ground.

Cyclic rate.—The rate of automatic fire, expressed in terms of shots per minute; especially, the maximum rate of fire of an automatic weapon.

Daily ration strength state.—*British term*: see **Daily telegram**.

Daily supply train.—*British term*: see **Daily train**.

Daily telegram.—A telegram or other message dispatched daily by divisions and larger units giving the unit's situation as regards supplies. A strength report is included. *British equivalent*: **Daily ration strength state**.

Daily train.—The train arriving daily at a railroad with supplies for troops which the railhead serves. *British equivalent*: **Daily supply train**.

Daily wastage rate.—*British term*: see **Day of supply**.

Danger area.—*a.* The area on a practice range where direct fire or fragments of bursting shells, ricocheting projectiles, etc., may cause casualties.

    b. British term: see **Danger space**.

Danger space.—That portion of the range within which a target of given dimensions would be hit by a projectile with a given angle of fall. *British equivalent*: **Danger area**.

Day of fire.—See **Unit of fire**.

Day of supply.—The estimated average expenditure of various items of supply per day in campaign expressed in quantities of specific items or in pounds per man per day. *British equivalent:* **Daily wastage rate**.

Day room.—A room in a barracks or company building set aside for soldiers' reading or recreation.

D day.—The day on which an operation is planned to begin. Subsequent days are referred to as D plus 1, D plus 2, and so forth. *British equivalent*: Same.

Dead abatis.—An obstacle turned toward the enemy made of cut-down or fallen trees, often interlaced with barbed wire. A *live abatis* is a similar barrier made of small trees or saplings bent down.

Dead area.—See **Dead space**.

Dead ground.—*British term:* see **Dead space**.

Deadline.—To remove from action, as for repairs. A tank or gun is deadlined for repairs.

Dead line.—A line in or around a military prison that a prisoner may not cross.

Dead reckoning.—Finding one's position by means of a compass and calculations based on speed, time elapsed, and direction from a know position. Dead reckoning is used for desert travel, coastwise shipping, and air navigation.

Dead space.—Ground which cannot be covered by fire from a position, because of intervening obstacles; dead area. *British equivalent:* ***Dead ground.***

Debarkation.—An unloading of troops, equipment, or supplies from a ship or aircraft.

Debouch.—(verb) (da BOOSH) *a.* To emerge from cover into an open area under enemy fire or ground observation. For example, "The rifle elements debouched from final assembly areas following the rear elements of the attacking light tanks." *British equivalent:* Same.

b. To emerge from a defile into a wider, more open area. For example, "The battalion debouched from the eastern exits of the village." *British equivalent*: Same.

Debouchment.—(noun) (da BOOSH munt) *a.* Emerging from cover into an open area under enemy fire or ground observation. *British equivalent:* Same.

b. Emerging from a defile into a wider, more open area. *British equivalent*: Same.

Decipher.—To translate a message in cipher from into plain text by the use of a cipher key.

Decision.—*a.* The general plan of a commander expressed definitely and briefly. *British equivalent*: ***Intention.***

b. A decisive outcome of a battle, one side being decisively defeated. *British equivalent*: Same.

Decode.—To translate a code message into ordinary language. *British equivalent:* Same.

Deliberate counter-attack.—*British term:* see ***Counterattack.***

Defend.—To maintain against force. To secure against attack. To conduct a defensive battle. *British equivalent*: Same.

Defended locality.—*British term*: see ***Defensive area*** and ***Defensive zone.***

Defense.—The means adopted for resisting attack. The act of defending, or state of being defended. *British equivalent*: ***Defence.***

Defense area.—That part of a battle position (or out-post zone) assigned to a unit as its area of responsibility in the all-around defense of the area of a higher unit, ordinarily used when referring to units smaller than a regiment. See also **Sector**. *British equivalent*: ***Defended locality.***

Defense forces.—Aviation assigned to provide close-in air defense of vulnerable and important areas to include, where necessary, reasonable protection against offshore carrier attacks. Compare ***Striking forces*** and ***Support forces***.

Defense in depth.—A system of mutually supporting positions organized for defense, with more powerful positions usually to the rear. Defense in depth is designed to break up and absorb the enemy attack.

Defense in place.—A system of defense based upon firm resistance without retreat, as opposed to delaying actions in successive positions.

Defense plan.—A coordinated plan for preventing or defeating an enemy. A defense plan includes plans for tactical organization, fire, security, air support, ground organization, counterattack, communications, and supply.

Defensive coastal area.—A part of a coastal zone and of the land and water adjacent to, and inshore of, the coast line within which defense operations will involve both Army and Navy forces. *British equivalent:* ***Coast defence area.***

Defensive fire.—*British term:* see ***Counterpreparation*** and ***Emergency counterpreparation*** and ***General counterpreparation*** and ***Local counterpreparation.***

Defensive obstacles.—*British term:* see ***Protective obstacles.***

Defensive-offensive.—The act of assuming the defensive with a view to permitting the enemy to exhaust his strength, and later to initiating an offensive in order to gain an objective. *British equivalent:* Same.

Defensive patrol method.—An antiaircraft defensive measure in which pursuit aviation is employed in the systematic search for and subsequent attack of enemy aircraft. *British equivalent:* ***Fighter patrol.***

Defensive position.—*a.* Any area occupied and more or less organized for defense. A battle position. A system of mutually supporting defensive areas or tactical localities of varying size. *British* *equivalent*: Same.

   b. Also British term: see **Battle position**.

Defensive sea area.—A portion of the coastal zone, usually including the approach to an important port, harbor, bay, or sound, within which, if such area be publicly proclaimed and neutrals notified, international practice tacitly permits the belligerent to extend his jurisdiction with a view to the protection of neutral shipping from mine fields, obstructions, or the danger of being considered hostile. *British equivalent*: None.

Defensive zone.—A belt of terrain, generally parallel to the front, which includes two or more organized or partially organized battle positions. *British equivalent*: **Defended locality**.

Deferred message.—A message whose delivery to the addressee may be delayed until the beginning of office hours of the morning following the day on which it is filed.

Defilade.—Protection from hostile ground observation and fire provided by a mask. Vertical distances by which a position is concealed from enemy observation. *British equivalent*: Same.

Defiladed.—Protected from hostile ground observation and fire, by a mask.

Defile.—A terrain feature or a structure which can be traversed only on a narrow front, or which restricts lateral movements; such as a mountain pass or a bridge. *British equivalent*: Same.

Deflagration.—An explosion that gradually builds up power smoothly and evenly, as opposed to **detonation**, which is practically instantaneous, and produces a sudden and violent explosion.

Delaying action.—A form of defensive action employed to slow up the enemy's advance and gain time without becoming decisively engaged, characteristic of the tactics of a rear guard in a retreat. Delaying action differs from **sustained defense**, the purpose of which is to stop an enemy attack at the defense line. *British equivalent*: Same.

Delaying position.—A position taken up for the purpose of slowing up or interfering with the advance of the enemy without becoming decisively engaged. *British equivalent*: **Intermediate position**.

Deliberate field fortification.—A fox hole, trench, gun emplacement, or obstacle constructed before contact with the enemy. A deliberate field fortification is generally more elaborate than a **hasty field fortification**, which is constructed under fire or threat of attack.

Deliberate fire.—Fire delivered more slowly than the normal fire, for the purpose of permitting precise adjustment and careful correction for greater accuracy of aim.

Delivery point.—*British term*: The point of transfer of loads to *first line transport*; the place where R.A.S.C. transport hands over its load to unit transport. See also *Meeting point.*

Demilitarize.—To take away all military organizations and installations; to restore civil government after a military occupation.

Demobilization.—Disbanding military forces; changing over from a war footing to a peacetime of inactive footing.

Demolition.—*a.* The act of demolishing; destruction, especially by explosives.

b. Used for demolition or destructive purposes; as, a demolition bomb.

Demonstration.—An attack delivered on a show of force made on a front where a decision is not sought and for the purpose of deceiving the enemy. *British equivalent:* Same.

Density.—The amount of traffic moving one way on a road, expressed in number of vehicles per mile. *British equivalent*: Same

Department.—One of the branches of the Army, such as the Adjutant General's Department or Judge Advocate General's Department, that carries out an administrative function of the Army.

Deploy.—To extend or widen the front of a military unit, extending from a close order to a battle formation. *British equivalent*: Same.

Deployment.—An extension of the front of a command. For example, to take up one of the formations prescribed in extended order. *British equivalent:* Same. (*Also British term: see* *Development*)

Depot.—An organized locality for the reception, classification, storage, issue, or salvage of supplies: or for the reception, classification, and forwarding of replacements. *Arm or service depots* pertain to a single arm or service and *general depots* pertain to two or more supply arms or services.

Depth.—The space from front to rear of any formation or of a position, including the front and rear elements. *British equivalent*: Same.

Desertion.—An unauthorized absence from a military post or duty with the intention of not returning, or of avoiding dangerous duty or important service. Desertion differs from *absence without leave*, in which the absentee does not have the intention of staying away or shirking dangerous duty.

Despatch.—*British term*: see **Dispatch** and **Message**.

Destruction fire.—Heavy artillery fire intended to cause the destruction of enemy works or matériel. Destruction fire is often used to put enemy artillery permanently out of action. *British equivalent*: **Destructive fire.**

Destructive fire.—*British term*: see **Destruction fire.**

Destructive shoot.—*British term*: see **Interdiction fire.**

Detach.—To take personnel or units away from the organization to which they belong, usually for special duty with another unit.

Detached post.—A post established outside the limits of the outpost proper for a special mission, as to observe or guard some locality of special importance,

Detachment.—*a.* A part of a unit separated from the main organization for duty elsewhere.

    b. A temporary military unit formed from other units or parts of units; a military unit that is a permanent separate unit smaller than an company.

Detail.—*a.* A soldier or soldiers assigned to a particular task, usually a temporary one, such as a guard detail.

    b. To assign a soldier or a group of soldiers to a particular task.

Detonate.—To explode suddenly and violently.

Detonating cord.—A flexible fabric tube containing a filler of high explosive that is set off by a blasting cap or by an electric detonator. It has an extremely high rate of explosion, and it is used to set off other high explosive charges or to act as a bursting charge. The detonating cord then in use was known commercially as **primacord.**

Detonator.—A sensitive explosive used in an explosive train.

Detraining point.—A point on a railroad where troops and equipment are unloaded.

Detrucking area.—An area that includes all the points in a given locality at which troops, their equipment, and supplies are unloaded from trucks.

Developing attack.—*a.* An attack made to obtain information about the strength and placement of enemy troops and the plans of action of the enemy.

    b. An attack made preliminary to a main attack in order to secure an advantage in position or timing that will make a position more secure or increase the probability of success of the main effort.

Development.—The distribution of a command from mass or route-column dispositions into smaller columns or groups in preparation for action. The extension in width and depth of companies and larger infantry units preparatory to approach march. *British equivalent*: **Deployment.**

Dictated order.—An order delivered orally. of which a verbatim record is made by the receiver. *British equivalent*: None.

Direct fire.—Fire in which the firer aims the weapon by means of sights directly at the target; fire by direct laying. Compare **Observed fire.**

Direction of march.—The direction in which the base of the command, whether actually in march or halted, is facing at the instant considered. *British equivalent*: Same.

Directive.—A military communication in which a policy is established or a specific action is ordered.

Direct laying.—Laying in which the sights of the weapon are aligned directly on the target. *British equivalent:* Same.

Direct support.—Support provided by that artillery which has the primary mission of supporting a designated subdivision of the combined force of which it is a part. In contrast to **general support**, which is given to the unit as a whole. *British equivalent: "In support."*

Discipline.—That mental attitude and state of training which render obedience and proper conduct habitual under all conditions. *British equivalent:* Same.

Disengage.—(verb) To break off action with an enemy.

Disengagement.—(noun) Breaking off action with an enemy.

Dispatch.—*a.* To send a message, order, report, or other communication.

   b. A message, order, report, or other communication that has been sent.

   c. To send or direct troops or other military units or conveyances.

   d. *British equivalent*: **Despatch** (in all meanings).

Dispensary.—A medical office where medicines are dispensed and dental treatment is furnished, without hospitalization. *British equivalent: Camp reception station.*

Displacement.—The movement of supporting weapons from one

75

firing position area to another. For example, in attack the successive movement of supporting weapons to correspond with the progress of the attacking echelon in order to keep weapons within efficient supporting distance thereof.

Disposition.—The distribution and the formation of the elements of a command and the duties assigned to each for the accomplishment of a common purpose. *British equivalent*: Same, also **Movement to new position.**

Distance.—The space from front to rear between men, animals, vehicles, or units in a formation. The space side to side is called an **interval**. *British equivalent*: Same.

Distant reconnaissance.—An exploration of objectives that lie outside the immediate striking range of a force, but about which detailed information is essential for military planning. The study of regions near at hand is called **close reconnaissance.**

Distributed fire.—Gunfire so placed in width that all parts of the target are under effective fire. Distributive fire may be used against a road, wood, or other large target; **concentrated fire** is directed at a single point or area.

Distributing point.—A place, other than a depot or railhead, where supplies are issued to regiments and smaller units. Distributing points are designated by the class of supplies therein, and by the identity of the unit establishing them, such as "Class I Distributing Point, 1st Division," or "Ammunition Distributing Point, 1st Infantry." *British equivalent*: **Ammunition point**, **supply point**, or **petrol point.**

Distribution.—The manner in which troops are disposed for any particular purpose, as battle, march, or maneuver. Dispersion of projectiles. An intentional dispersion of fire for the purpose of covering a desired frontage or depth, accomplished in various ways. A delivery of supplies, specifically by the supply officer of a higher unit to subordinate units, or to individuals.

Distribution, dump (railroad) (unit).—See **Dump (railroad) (unit) distribution.**

District court-martial.—*British term*: A court martial made up of at least three officers for noncapital crimes. It cannot try officers, or persons subject to military law as officers. It can try a warrant officer, but its powers of punishing him are limited. See **Special court-martial.**

Dive bomber.—A bomber that releases its bomb load just before it pulls out of a dive toward a target. The angle is such that the pilot sights through his gun sights.

Diversion.—An attack or sham attack intended to hold the enemy's attention and draw his troops from the point at which a main attack is to be made; an attack upon a weak point in the enemy's battle line intended to make him draw some his forces away from an attack at another point.

Division.—a. A major administrative and tactical unit. A division is larger than a regiment or brigade, and small than a corps. It is usually commanded by a major general.

b. A branch or section of the headquarters of a division or higher unit, that handles military matters of a particular nature, such as personnel, intelligence, plans and training, or supply and evacuation.

Divisional administrative area.—*British term*: An area in which the administrative units and unit B echelon vehicles, whose functions do not require them to be situated elsewhere, are situated. See **Rear echelon.**

Divisional administrative group.—*British term*: The administrative units and unit B echelon in the administrative area. See **Rear echelon.**

Division artillery.—Artillery that is permanently an integral part of a division. For tactical purposes, all artillery place under the command of a division is considered division artillery.

Division engineer.—A senior engineer officer in command of all engineer troops of a division. He is a member of the special staff of the division commander.

Division trains.—All the service elements of a division, including the maintenance battalion, supply battalion, medical battalion, etc. They are organized under the train headquarters.

Dock.—A slip or waterway, as between two piers, for the reception of ships. *British equivalent*: Same.

Dogfight.—Combat between individual planes or between individual mechanized units.

Domination of the objective.—*British term*: To remain on the objective in order to neutralise the enemy opposition. Tanks frequently neutralise the enemy position by keeping the objective under effective fire without themselves actually being upon it.

Double action.—A method of firing in a revolver and in old-style rifles and shotguns in which a single pull of the trigger both cocks and fires the weapon, in contrast to **single action**, in which the hammer must be cocked by hand before firing.

Double envelopment.—An attack on both flanks of the enemy while his center is held in check.

Draft.—*a.* A selection of persons from the total manpower of a country for compulsory military service. In this meaning, in the United States also called **Selective Service**.

b. Persons selected for compulsory military service.

c. To call a man for compulsory military service.

d. A request for the delivery of supplies covered by credit. In this meaning, also called **call**.

Draftee.—See **Selectee**.

Dress.—The act of taking a correct alinement.

Dressing station.—See **Aid station**.

Drill.—The exercises and evolutions taught on the drill ground and practiced for the purpose of instilling discipline, control, and flexibility.

Dropping ground.—A place where messages are dropped to ground troops from airplanes.

Dropping zone.—*British term*: An area on to which parachute troops land or where stores are delivered to a formation by dropping. See **Jump area** and **Landing area**.

Drop message.—A message dropped from an airplane to a ground unit. Maps and photographs, which cannot be transmitted by radio, are often delivered as drop messages.

Dual status.—The holding of two appointments by a single officer in two different military organizations at the same time, particularly in the National Guard and in the National Guard of the United States.

Dugout.—An underground shelter built to protect troops, ammunition, and matériel from gunfire.

Dump.—A temporary stockage of supplies established by a corps, division, or smaller unit. When supplies are ordered and issued from dumps, the latter become distributing points. Dumps are designated by the identity of the unit establishing them and the class of supplies therein, such as "1st Infantry Ammunition Dump" or "1st Division Class I Supply Dump." *British equivalent:* Same.

Dump distribution.—Issue of Class I or other supplies to regimental or (similar unit) transportation at a dump established by higher authority.

Dynamite.—*a.* A powerful explosive made from a nitroglycerin base. It is highly sensitive to shock, but less powerful than trinitrotoluene (TNT).

b. To blow up or to destroy with dynamite.

Earthwork.—A field fortification made chiefly of earth, such as a trench.

Echelon.—*a.* A formation in which the subdivisions are placed one behind another extending beyond and unmasking one another wholly or in part; *British definition*: A formation of successive and parallel units facing in the same direction, each on a flank and to the rear of the unit in front of it.

b. In battle formation, the different fractions of a command in the direction of depth, to each of which a principal combat mission is assigned, such as the attacking echelon, support echelon, and reserve echelon. *British equivalent:* Used only for combined infantry and tank attacks.

c. The various subdivisions of a headquarters, such as ***forward echelon*** and ***rear echelon***.

d. British (Organizational).—Refers only to transport. The division of a particular organization into front and rear parts, e.g., A and B echelons of first-line transport.

Echelon maintenance.—A system of maintenance and repair of matériel and equipment in which jobs are allocated to organizations in accordance with the availability of personnel, tools, supplies, and time within the organization.

First echelon maintenance.—The servicing or repairs that can be done by an operator, driver, or crew.

Second echelon maintenance.—The servicing or maintenance that is beyond the scope of the operating personnel, but which can be done by the maintenance section of a unit that uses the equipment.

Third echelon maintenance.—The maintenance, repairs, and unit replacement beyond the scope of the troops using the matériel and equipment, which can be performed by mobile maintenance organizations.

Fourth echelon maintenance.—The general overhaul and reclamation of equipment, units, and parts, involving the use of heavy tools and the services of general and technical mechanics.

Fifth echelon maintenance.—The maintenance of equipment by personnel of maintenance and supply units located at fixed installations in the rear areas. This includes the reclamation of matériel, the limited manufacture of parts and equipment, and the supplying of equipment to lower echelons.

Echelonment.—Forming or being formed into echelons.

Echelon of attack.—See ***Attack (or attacking) echelon.***

Economy of force.—A fundamental principle of warfare that only a minimum of troops and supplies should be used on less important objectives so that the main strength can be reserved for a major effort.

Effective beaten zone.—A section of a target area in which a high percentage of shots fall, usually the zone that gets 85 percent of the hits; effective pattern eighty-five percent zone.

Effective range.—The range at which, for a particular gun, effective results may be expected. The distance at which a gun may be expected to fire accurately to cause casualties or damage. The ordinary effective range of an automatic pistol is 25 yards; its maximum effective range is 75 yards; its extreme range, if held at an angle of 30 degrees, is about 1600 yards. *British equivalent:* Same.

Effectives.—Troops actually available for combat duty.

Ejector.—A mechanism in small arms and rapid-fire guns which automatically throws an empty cartridge case or unfired cartridge from the breech or receiver. The ***ejector*** is not to be confused with the ***extractor***, which pulls the empty cartridge case or unfired cartridge from the chamber.

Element.—One of the subdivisions of a command. The term "elements" is used in an inclusive sense to refer to all those various smaller units or parts of units, generally different in character, as service elements, meaning quartermaster, ordnance, engineer, and medical units, etc. *British equivalent:* Same.

Elevation.—The vertical angle between the line from the muzzle of the gun to the target and the axis of the bore when the gun is pointed for range. Also called ***angle of elevation.***

Embargo.—A ruling given by the Chief of Transportation, or other authority, forbidding truck or ship movements to or from a given place. Embargoes are put into effect when traffic becomes congested, and may be disregarded only by special permission.

Embarkation.—The loading of troops, equipment, or supplies on to a ship or aircraft.

Embrasure.—An opening in a wall or parapet, especially one through which a gun is fired. It is usually cut wider at the outside to permit the gun to swing through a greater arc.

Embus.—*British term*: see **Entruck.**

Embussed movement.—*British term*: see **Troop movement by motor.**

Embussing (debussing) point.—*British term*: see **Entrucking (detrucking) point.**

Emergency barrage.—A barrage which may be ordered fired to cover gaps in the normal barrage line or to reinforce the normal barrage of another part of the line. *British equivalent:* **Superimposed fire.**

Emergency counterpreparation.—Fire planned by the artillery of one division to reinforce the local counterpreparations of other divisions. *British equivalent:* **Defensive fire** (though normally applied only to own front).

Emergency landing field.—A place where aircraft can land in an emergency. It usually does not have facilities for shelter, supply, and repair.

Emplacement.—A prepared position from which a unit or a weapon executes its fire missions. See **Firing position.** *British equivalent:* Same.

Employment reconnaissance.—A reconnaissance, before combat, by an armored unit, for the purpose of finding out the best use that can be made of an armored force in a tactical situation. Particular attention is given to routes, observation points, covers, and the character of the terrain.

Encampment.—A temporary camp in the field.

Encirclement.—Surrounding or hemming in a hostile force being surrounded in a double-flank maneuver.

Encircling force.—A pursuing force which moves around the hostile flanks or through a breach to reach and attack the heads of retreating enemy columns and bring them to a halt; a force that tries to surround the enemy in order to destroy his communication and supply lines and to cut off his line of retreat. *British equivalent:* **Enveloping force.**

Encode.—Translate ordinary language into code. To prepare a message in code. *British equivalent:* Same.

Encounter (or contact) battle.—*British term*: see **Meeting engage-**

ment.

Encrypt.—Translate a message from plan text into secret code or cipher. Also called ***cryptograph.***

Endurance.—The maximum time, usually at a given speed and altitude, that an aircraft can stay in the air without refueling.

Enemy capabilities.—Lines of action the enemy can take in a given tactical situation.

Enfield rifle.—A popular name for the United States rifle, caliber .30, model 1917. It is a bolt-type, breech-loading magazine rifle.

Enfilade.—*a.* (verb) To fire at a target so that the fire coincides with the long axis of the target. For example, to fire against troops disposed in a generally linear formation from their direct flank and along the direction of their front. *British equivalent*: Same
   b. (noun) Flanking or frontal fire which sweeps along the length of a target. In this meaning, also called ***enfilade fire.*** *British equivalent*: Same.

Enlisted cadre.—The key enlisted men needed to organize and train a new unit.

Enlisted man.—A noncommissioned officer or private; any member of the Army who is below the grade of a commissioned officer or warrant officer. *British equivalent*: ***Ratings*** (Royal Navy), ***Other ranks*** (British Army, includes warrant officers), or ***Airman*** (Royal Air Force, includes warrant officers).

Enlistment.—*a.* A voluntary enrollment as a member of the Army, as contrasted with induction under the draft of Selective Service.
   b. A period of military service under the contract of enlistment.

Ensign.—*a.* A flag or banner; especially, a national flag or banner.
   b. The lowest commissioned officer the Navy, who ranks next below lieutenant, junior grade. An ensign in the Navy has a rank equivalent to that of second lieutenant in the Army.

Entrain.—To get on a train; to put men on a train. Entraining is usually connected with troop movement by rail.

Entraining point.—A point on a railroad at which troops or vehicles are loaded on trains.

Entrench.—See ***Intrench.***

Entrenchment.—See ***Intrenchment.***

Entruck.—To get in a truck, to put men in a truck. Entrucking is usually connected with troop movement by truck. *British equivalent*:

Embus.

Entrucking group.—Troops, matériel, or supplies properly disposed for loading at an entrucking point. *British equivalent*: None.

Entrucking (detrucking) point.—The point at which the head of a truck column halts for the entrucking (detrucking) of troops or supplies. *British equivalent: **Embussing (debussing) point.***

Envelop.—To attack one or both flanks of the enemy, usually attacking his front at the same time.

Enveloping force.—*British term*: see ***Encircling force.***

Envelopment.—An offensive maneuver in which the main attack is directed from an area wholly or partially outside and to the flank(s) of the initial disposition of the enemy's main forces and toward an objective in his rear; usually assisted by a secondary attack directed against the enemy's front. If the attack is made on both flanks at once it is called a ***double envelopment***. *British equivalent*: Same.

Equipage.—All supplies necessary for a man or a unit in the field, excepting arms and clothing.

Equipment.—All articles needed to outfit an individual or organization. It includes clothing, tools, utensils, weapons, and supplies.

Equipment (War).—*British term:* The approved "holding" of ***stores*** by a unit.

Escarpment.—A steep, cliff-like slope cut out of the ground as a fortification against enemy attack.

Escort.—A body of armed men to guard a person, persons, or goods on a journey, or to accompany as a mark of respect or honor. *British equivalent:* Same.

Escort force.—A part of the naval local defense forces charged with the duty of protecting convoys within naval district waters. *British equivalent:* Same.

Espionage.—The process of obtaining information of the enemy by means of spies.

Essential elements of information.—That information of the enemy, of the terrain not under control, or of meteorological conditions in territory held by the enemy, which a commander needs in order to make a sound decision, conduct a maneuver, avoid surprise, or formulate the details of a plan. They include questions relating to enemy capabilities, other intelligence specifically desired by the commander, and information requested by other units. *British equivalent:* Same.

Estimate of the situation.—A logical process of reasoning by which a commander considers all available data affecting the military situation and arrives at a decision as to a course of action, including the expression of his decision. *British equivalent:* **Appreciation of the situation.**

Evacuation.—The withdrawal of troops or civilians from a given area; also, the act of clearing personnel (such as stragglers, prisoners of war, sick, and wounded), animals, or matériel (such as salvage and surplus baggage) from a given locality. *British equivalent:* Same.

Evacuation hospital.—A mobile field hospital in the combat zone. It gives necessary treatment to casualties, but sends serious cases on to fixed hospitals for further care.

Evaluation of information.—An analysis of information to determine its probable intelligence value; i.e., its accuracy, its credibility, and its application to the situation. *British equivalent:* **Assessment of value of intelligence report.**

Evolution.—A maneuver or movement by which a command changes from one position or formation to another.

Exchange.—A military organization that sells merchandise and services to military personnel and other authorized personnel. Often called an *Army exchange* or *post exchange.*

Executive officer.—The principal assistant of the commander; executive. The executive officer is charged with supervising the work of the staff in a command not provided with a General Staff; generally, the second-in-command. *British equivalent:* No exact equivalent (the Adjutant, and not the second-in-command, performs these duties in the battalion or equivalent unit).

Expeditionary force.—An armed force for foreign service; especially, large forces to invade, or fight in, other countries.

Exploitation.—*a.* The act of taking full advantage of success in battle and following up initial gains. See also **Attacking, continuing.** *British equivalent:* Same.

b. Taking full advantage of any information that has come to hand; examination of the information obtained from photographic or other sources, for tactical or strategic purposes.

Explosive.—Any chemical compound or mixture which burns so rapidly that an explosion results. A **high explosive** is used as a bursting charge in bombs or projectiles; a **low explosive** is used as a propelling charge in guns or for ordinary blasting.

Exposed.—Unprotected from enemy attack. An **exposed flank** is a side of a defended position or formation that has been left unprotected from enemy attack. An **exposed position** is a defended position that has been left unprotected from enemy attack.

Extended formation.—A formation whose elements are wide apart but still within sight and actual fire support of each other. Extended formations are the formations normally used in combat.

Extended order.—Formations in which the individuals or elements are separated by intervals or distances, or both, greater than in close order.

Extractor.—*a.* A device for pulling an empty cartridge or an unfired cartridge out of the chamber of a gun. The extractor is not to be confused with the ejector, which throws the empty cartridge case or unfired cartridge out of the receiver. In this meaning, also called **shell extractor.**

b. A device in certain types of automatic weapons which pulls the round from a feed belt.

Extreme range—The maximum range of any weapon. *British equivalent:* Same.

Fascine.—A long bundle of sticks, poles, or rods tied together. Fascines are used as trail supports for guns, and in making retaining walls, in filling ditches, and in similar field construction.

Feed belt.—A fabric or metal ammunition band with loops for cartridges that are fed from it into a machine gun or other automatic weapon. Also called **ammunition belt.**

Feint.—An attack or demonstration intended to deceive the enemy. A pretense. A stratagem. To make a feint. *British equivalent:* Same.

Field army.—See **Army.**

Field artillery.—Artillery mounted on carriages and mobile enough to accompany infantry, cavalry, or armored units in the field.

Field bag.—See **Musette bag.**

Field cap.—See **Garrison cap.**

Field expedients.—Improvised means used to facilitate continued employment of equipment in the field. Particularly applicable to automotive equipment.

Field fortification.—The act of increasing the natural strength of a defensive position by words designed to permit the fullest possible fire and movement of the defender, and to restrict to the greatest possible extent the movement and the effects of the fire of the attacker. Defensive works of a temporary nature used in the field in both the attack and defense.

Field general court-martial.—*British term*: A court-martial that may be convened by any officer commanding a detachment or portion of troops abroad, or the commanding officer of any corps or portion of a corps on active service, or any officer in immediate command of a body of forces on active duty. A field general court-martial has the same powers as a general court-martial, provided that the court is composed of at least three officers. If, in the opinion of the convening officer, three officers are not available two officers are legally sufficient, but a court consisting of two officers cannot award a sentence in excess of imprisonment.

Field grade.—A classification of officers above a captain and below a brigadier general. Field grade includes colonels, lieutenant colonels, and majors.

Field gun.—A field artillery piece; a cannon mounted on a carriage for use in the field.

Field hospital.—A mobile hospital which may be divided into units and employed in the field under tentage or other locally improvised shelter.

Field marshal.—A rank in most European armies that is equivalent to a General of the Armies.

Field officer.—An officer who ranks above a captain and below a brigadier general, that is, between a general officer and a company officer. Colonels, lieutenant colonels, and majors are field officers. *British equivalent*: Same.

Field of fire.—An area that a gun or battery covers effectively. *British equivalent*: Same.

Field order.—An order conveying the directions of the commander to the subordinate commanders charged with the execution of tactical operations. *British equivalent*: **Operation order**.

Field ration.—A ration that is prescribed for use only in time of war or national emergency when the garrison ration is not issued. It is issued only in actual articles, not money, and consists of the following:

Field ration A corresponds in general to the peacetime garrison ration and is generally perishable; it is not suitable as a reserve ration.

Field ration B is the same as field ration A except that nonperishable substitutes replace perishable items. This ration is suitable for reserve purposes.

Field ration C is a cooked balanced ration in cans. Each ration consists of three cans of prepared meats and vegetables and three cans of crackers, and soluble coffee. As this ration is not perishable, it is suitable for use as a unit reserve or and an individual reserve.

Field ration D consists of three 4-ounce chocolate bars per ration. It is a nonperishable ration and is suitable for use as an individual reserve.

Field train.—Formerly, the train of a regiment or similar unit carrying unit reserves of rations, forage, fuel, and organization equipment and baggage not needed initially in combat. See *Train. British equivalent: "B" echelon transport.*

Field Transport, R.A.S.C.—*British term*: It is divided into:

Third line transport.—Consists of ammunition parks and petrol (gasoline) parks which work normally between *railhead* and *refilling points.*

Second line transport.—Consists of ammunition companies and petrol (gasoline) companies which work normally between *refilling points* and *delivery points*. In each petrol company there is a section for the carriage of blankets and reserve clothing.

Supply columns.—These columns carry supplies, mails, engineer and ordnance stores and work between *railheads* and *delivery points*. Each column is divided into two similar echelons which deliver supplies on alternate days.

Miscellaneous.—Other units such as troop carrying companies, motor ambulance convoys, and reserve motor transport companies

Fifth echelon maintenance.—Maintenance of equipment by personnel of maintenance and supply units located at fixed installations in the rear areas. This includes the reclamation or complete reconditioning of matériel, the limited manufacture of parts and equipment, and the supplying of equipment to lower echelons.

Fighter airplane (or aircraft).—An aircraft used to seek out and destroy enemy aircraft in the air. A fighter airplane has high speed, a high rate of climb, and great maneuverability, but relatively short range. It is used as accompanying support for other aviation, for interception and pursuit in general air defense, and sometimes for attacks on light surface objectives. See also *Pursuit aviation.*

Fighter-bomber.—An airplane that combines the function of a fighter airplane with that of a bomber.

Fighter command.—A tactical and administrative unit of the U.S.

Army Air Forces that is concerned primarily with breaking up enemy air attack and supporting air or ground striking forces. A fighter command is larger than a wing and smaller than an air force.

Fighter Command.—*British term*: A subdivision of the Royal Air Force responsible for the air defense of the United Kingdom. See also **Air defense command.**

Fighting (or "F") group.—*British term*: *see* **Combat echelon** and **Group.**

Fighting troops.—*British term*: Cavalry, artillery, engineers, signals, infantry and tank corps, and any air force contingent co-operating.

File closer.—An officer or noncommissioned officer placed in rear of a rank to supervise the men in ranks and see that the orders of the leader are carried out. For convenience, this term is applied to any man posted in the line of file closers.

Filler replacement.—An officer or enlisted man assigned to a newly organized unit to bring it to its prescribed strength. A filler replacement differs from a loss replacement, who is an officer or enlisted man who takes the place of a person killed, wounded, or lost from other causes. *British equivalent*: **First-line reinforcement.**

Final assembly area.—A place where troops are concentrated in preparation for an attack. Troop units have been organized in various **initial assembly areas** are moved up to the final assembly area, from which they move into combat.

Final protective fire.—Concentrated fire along a line close to a position, where the last and strongest defense would have to be made. Final protective fire may be fire from machine guns along their final protective lines, from mortars onto their primary target areas, and from artillery onto their normal barrage lines.

Final protective line.—For machine gun fire, a predetermined line along which, in order to stop assaults, is placed grazing fire, often fixed as to direction and elevation, and capable of delivery under any condition of visibility. *British equivalent*: **Fixed line.**

Fire adjustment.—Correcting the elevation and left and right direction of a gun, or regulating the explosion time of its projectile, so that the projectile will strike or burst at the desired point.

Fire and movement.—A method of attack in which the advancing element is supported and covered by gunfire of other elements; fire and maneuver. *British*: Same.

Firearm.—A gun from which projectiles are fired. All sizes of guns are firearms, but firearms usually means small arms, which include rifles, pistols, and other guns that a man may carry.

Fire, assault.—Fire delivered by the unit while advancing at a walk. *British equivalent*: **Fire on the move**.

Fire bay.—*British term*: A length of trench from which it is intended to deliver rifle fire. See also **Firing bay.**

Fire, collective.—The combined fire of a group of individuals. *British equivalent*: Same.

Fire, combined traversing and searching.—Fire distributed both in width and depth by changes in direction and elevation of the gun.

Fire, conduct of.—Employment of technical means to place accurate fire on a target. *British equivalent*: **Fire control**.

Fire control.—Fire control includes in its scope all operations connected with the preparation and actual application of fire to a target. Specifically, it includes supervision over the computation of fire data, the designation and engagement of targets, class and rate of fire, the opening and cessation, and number of rounds. Fire control is exercised by commanders having under their command one fire unit or several fire units controlled as a unit. For example, a machine-gun platoon of two sections may fire as a unit at the command of the platoon leader (fire control) or by the assignment of fire missions (sectors of fire, target areas) to the sections (fire direction). *British equivalent*: Same. (*Also British term: see* **Fire, conduct of**)

Fire, converging.—Fire from different directions brought to bear upon a single point or area. *British equivalent*: Same.

Fire, destruction.—Artillery fire delivered for the express purpose of destruction and when it is reasonable to expect that relatively complete destruction can be attained. *British equivalent*: **Destructive fire**.

Fire, direct.—Fire in which the sights of the weapon are aligned directly on the target. *British equivalent*: Same.

Fire, direct laying.—*British term*: When a gun is laid by looking over or through the sights at the target.

Fire direction.—A commander exercises fire direction functions by the assignment of fire missions to the component fire units of his command. Instructions for fire direction include the designation of target areas or sectors of fire where fire is to be placed; they may also prescribe the conditions under which fire is to be opened (hour, order of fires) and the ammunition expenditure for each fire. Fire direction is exercised by commanders having several fire units under their command. *British equivalent*: Same.

Fire discipline.—Fire discipline is a state of order, coolness, efficiency, and obedience existing among troops engaged in a fire fight, The enforcement of fire discipline is the function of leaders in immediate command of the soldiers engaged in the operation of weapons and the delivery of fire. *British equivalent*: Same.

Fire distributed.—Fire distributed in width for the purpose of keeping all parts of the target under effective fire. *British equivalent*: Same.

Fire, field of.—The area in the direction of the enemy which can be effectively covered by the fire of a firing unit, from a given position. That portion of the terrain or water area covered by the fire of a gun, battery, or other unit. *British equivalent*: Same.

Fire fight.—A delivery of fire between opposing infantry or gun units. Fire fight is a phase of an attack that usually follows the approach march and deployment and comes before the assault.

Fire, fixed (concentrated) (point).—Fire directed at a single point, without traversing or searching. *British equivalent:* **Fire on fixed line**.

Fire, flanking.—Fire directed against a unit or objective from an area on its flank. Flanking fire may be enfilade or oblique. *British equivalent:* Same.

Fire for adjustment.—Fire delivered primarily for the purpose of correcting, by observation, inaccuracies in the firing data. *British equivalent*: **Registration** or **trial shoot**.

Fire for effect.—Fire delivered for the purpose of neutralizing or destroying a target, or the accomplishment of the tactical effect sought. Any fire against a hostile target, other than for registration. *British equivalent*: Same.

Fire, frontal.—Fire delivered approximately at right angles to the front of the enemy's line, or other linear target. *British equivalent*: Same.

Fire, grazing.—Fire which is approximately parallel to the surface of the ground and does not rise higher above it than the height of a man standing. Fire with a long or continuous danger space. *British equivalent:* Same.

Fire, high-angle.—*a.* Fire delivered at elevations greater than the elevation corresponding to the maximum range.

b. British: Fire from all guns and howitzers at all angles of elevation exceeding 25°.

Fire, indirect.—Fire in which the weapon is aimed by indirect laying.

Fire, indirect laying.—*British term*: When a gun is laid for direction on an aiming point or on aiming points and elevation adjusted by sight clinometer.

Fire, leading.—Fire delivered to strike a moving target. *British equivalent*: ***Aiming off***.

Fire, low-angle.—Fire delivered at angles of elevation below that required for maximum range. *British equivalent*: Same.

Fire mission.—A specific assignment given to a fire unit as part of a definite plan. A fire mission includes the assignment of targets and full directions for timing of fire and for the guns to be used.

Fire, oblique.—Fire delivered from a direction oblique to the long axis of the target. *British equivalent*: Same.

Fire, observed.—Fire which is adjusted by observation, especially from an observation post at, or in communication with, the gun position. *British equivalent:* Same.

Fire on targets of opportunity.—Fire on targets appearing suddenly or unexpectedly during the course of an engagement. *British equivalent: **Fire on opportunity (or gun fire) targets***.

Fire on the move.—*British term*: see ***Fire, assault***.

Fire order.—A command that directs and controls the fire of a unit, gun, or group of guns in accordance with the plan of the commander.

Fire, overhead.—Fire that is delivered over the heads of friendly troops. *British equivalent*: Same.

Fire plan.—A tactical plan for using the weapons of a unit so that their fire missions will be coordinated. The fire plan includes the assignment of fire missions to weapons and instructions for timing of fire, signals, shifts in position, etc.

Fire, plunging.—Fire in which the angle of fall of the bullets with reference to the slope of the ground is such that the danger space is practically confined to the beaten zone and the length of the beaten zone is materially lessened. *British equivalent*: Same.

Fire (or firing) position.—A locality or emplacement from which a unit or a weapon executes fire missions; classified as primary, alternate, or supplementary (see definitions under respective titles).

Fire, searching.—Fire distributed in the direction of depth by successive changes in the elevation of the gun. *British equivalent:* Same.

Fire, searching and sweeping.—*British term:* see **Fire, combined traversing and searching.**

Fire superiority.—A condition of fire whose effect is greater than that of the enemy because of its greater accuracy and volume. Fire superiority makes possible advances against the enemy without heavy losses.

Fire support.—Using the fire of various weapons according to a tactical plan to assist infantry and other units.

Fire, traversing.—Fire distributed in the direction of width by successive changes in the direction of the gun. *British equivalent:* **Sweeping** (or **traversing**) **fire.**

Fire trench.—Trench designed primarily to provide cover for personnel when delivering fire. *British equivalent:* Same.

Fire unit.—A unit whose fire in battle is under the immediate and effective control of one leader. The rifle squad is one fire unit in the infantry; the battery is the fire unit in the artillery. *British equivalent:* Same.

Firing battery.—That part of a battery actually at the firing position when a battery is prepared for action. It includes the pieces, personnel, and equipment necessary for their operation

Firing bay.—One of a series of short straight sections of a fire trench, set forward and joined to the next by short trenches which make an indentation in any of various shapes; fire bay. A fire trench is divided into firing bays so that a bomb or shell falling in one bay does not cause destruction in those on either side. *British equivalent:* **Fire bay.**

First echelon maintenance.—The servicing or repairs that can be done by an operator, driver, or crew.

First flight.—*British term:* In combined operations, troops conveyed in the landing craft making the first trip to the beaches.

First lieutenant.—An officer in the Army who ranks next above a second lieutenant and next below a captain. A first lieutenant is usually second in command of a company.

First-line reinforcement.—*British term:* see **Filler replacement.**

First line transport.—*British term:* see **Transport (British)**.

First sergeant.—The chief noncommissioned officer in a company, battery, or similar unit in the Army. He is in immediate charge of the enlisted men of the unit, and holds a rank equal to that of a master sergeant.

Fishbone.—A series of independent underground passages that military engineers cut out in the direction of the enemy with branches for purposes of attack, flank protection, and listening. A fishbone differs from a *lateral*, which is an underground passage cut parallel to the front line, form which galleries are carried to the enemy.

Fix.—*a.* A point on a map or chart at which two lines of position cross one another, used for finding the position of an aircraft, ship, etc.
   b. To stop an enemy and hold him where he is.

Fix bayonets.—A command to attach bayonets to rifles in readiness for use.

Fixed ammunition.—Ammunition that is loaded into a weapon as a unit, and not in parts, as in **separate-loading ammunition**. The cartridge case is attached to the projectile in fixed ammunition.

Fixed armament.—Seacoast artillery weapons that are emplaced in permanent firing positions. *British equivalent:* Same.

Fixed line.—*British term*: see **Final protective line**.

Fixed obstacles.—Obstacles which are securely placed or fastened. *British equivalent:* Same.

Flag.—A piece of cloth with a color or design that has a special meaning or serves as a signal. *Examples*: the flag of the United States, the white flag of truce, and weather flags to let people know what kind of weather is coming. In military service, the **color** is a flag carried by dismounted units, an **ensign** is a national flag, a **pennant** is a small triangular flag usually flown for identification of a unit or a general officer, a **standard** is a flag carried by mounted or motorized units, and a **guidon** is a flag carried by Army units for identification.

Flank.—The side of a command from the leading to the rearmost element, inclusive. **Right** flank is the right side when facing the enemy, and does not change when the command is moving to the rear. *British equivalent*: Same.

Flank guard (or patrol).—A security detachment designed to protect the flank of a marching force. *British equivalent*: Same.

Flanking attack.—Attack directed against the flank of a hostile force.

British equivalent: Same.

Flanking fire.—Fire directed against a unit or objective from an area on its flank. Flanking fire may be enfilade or oblique.

Flank march.—A march across or nearly parallel to the front of the enemy.

Flank security.—Measures for the protection of the flanks of a marching column or of a force in battle formation. Whenever practicable, flank security employs flank guards or patrols, and natural and artificial barriers.

Flash message.—A brief message in special form sent by telephone, telegraph, or radio to give information about enemy aircraft. A flash message has priority over all other messages.

Flash ranging.—Finding the position of the burst of a projectile or of an enemy gun by observing its flash.

Flat trajectory.—A trajectory with little vertical curvature. Trajectories fired from long guns with high muzzle velocity are comparatively flat, as are those fired at short range.

Flight.—The basic tactical unit of Army Air Forces, consisting of two or more aircraft. *British equivalent:* Same.

Floating reserve.—*British term:* In combined operations, that portion of the *covering force* retained in reserve, embarked on ships or landing craft and ready for immediate landing when and where required.

Flying boat.—An airplane that can float on water. A flying boat is a seaplane with a boat-shaped hull that is both the fuselage of the plane and the means by which the craft is supported on water.

Flying ferry.—A raft, used as a ferry, that is controlled in the stream by an anchor cable fastened farther upstream; flying bridge. Usually the current moves the flying ferry from shore to shore, but sometimes outboard motors are also used.

Flying officer.—An Army Air Forces officer who holds a rating as a pilot, aircraft observer, or other member of a combat crew of a military aircraft. In time of war, flight surgeons and also officers and warrant officers undergoing flying training are classified as flying officers.

Follow.—To regular movement on the element in front. *British equivalent:* Same.

Follow-up.—The act of exerting close direct pressure on a withdrawing force. *British equivalent:* Same.

Forage.—Food for animals. To collect supplies for men and animals. *British equivalent:* Same.

Foragers.—Mounted troopers abreast of each other with intervals greater than those prescribed for close order. *British equivalent:* None.

Forced march.—Any march in which the march capacity of foot and mounted troops is increased by increasing the number of marching hours per day rather than by increasing the hourly rate of march. *British equivalent:* Same.

Forces in the field.—*British term:* The whole of the forces in the theater of operations, whether naval, military, or air, subject to the command of the C.-in-C.

Foreshore.—*British term:* Used in combined operations, that part of the beach which lies between the high and low water marks.

Fork.—A change in range elevation or direction necessary to move the center of impact of artillery fire the distance of four *probable errors.* Fork is sometimes used as a unit of change in the conduct of fire.

Formation.—*a.* The arrangement of the subdivisions of a command so that all elements are placed in order in line, in column, in echelon, or in any other designated disposition. *British equivalent:* Same or *forming up.*

b. British equivalent: A combination of units of different arms and services of a brigade or more.

Forming up.—*British term:* see **Assembly** and **Formation**.

Forming-up place.—*British term:* a. In night operations, the position where troops detailed for the assault will deploy.

b. In river crossings, places at which folding boat equipment is opened and carrying, rowing, and assaulting parties are formed into their final order for the assault, etc. When Kapok bridging equipment is used such points are termed **Bridge forming points**.

c. In combined operations, a place of assembly for smaller units, clear of, but close to, the landing-place, to which troops proceed immediately they have landed.

Forward ammunition point.—*British term:* The advanced point established by ammunition companies to facilitate supply of ammunition to units.

Fort.—A land area within a harbor defense wherein are located harbor defense elements. A permanent post.

Fortress command.—*British term*: see **Coast artillery district.**

Forward area.—A section of a theater of operations in which attack by enemy ground forces is probable. The forward area includes, primarily, the area covered by the combat zone. (*Also British term*: see **Combat zone.**)

Forward base.—*British term*: When the distance between the forward formations and the base, or advanced base, becomes too great to guarantee punctual delivery, it may be necessary to establish fore holdings and installations. See also: **Main base.**

Forward (foremost) defended localities.—*British term*: see **Main line of resistance.**

Forward echelon.—A station of a unit's headquarters where the commander and staff work. In combat, a unit's headquarters is often divided into a **forward** and a **rear echelon**; the forward echelon is called the **command post.**

Forward HQ.—*British term*: see **Command post.**

Forward maintenance area.—*British term*: An area formed, when considered necessary, into which Royal Army Service Corps transport under army control empties, and from which second line transport draws, and at which small stocks are held.

Forward observation post.—An observation station set up ahead of the main battle position.

Forward observer.—An observer at a forward observation post, especially for the purpose of adjusting artillery fire. A forward observer may be a roving observer or may be at a fixed post, depending on the military situation.

Forward standing patrol.—*British term*: see **Outguard.**

Fourragère.—(FOOR a ZHAYR). A metal-tipped, braided cord worn around the left shoulder. A fourragère is given to anyone who has a required number of citations, or to all members of a unit decorated for conspicuous bravery in action. When a unit has been so decorated, future as well as present members of the unit wear the fourragère.

Fox hole.—Small, individual shelter or rifle pit. *See* **Shelter trenches.** *British equivalent*: **Slit trench** (that permits firing from a standing position) or **weapon pit.**

Fragmentary orders.—Combat orders issued in fragmentary form, and consisting of separate instructions to one or more subordinate units prescribing the part each is to play in the operation or in the separate phases thereof. *British equivalent:* **Operation instructions**.

Fragmentation.—The breaking up and scattering of the fragments of a shell, bomb, or grenade. *British equivalent:* Same.

Fragmentation bomb.—A bomb intended primarily for use against personnel on the ground.   See also **Antipersonnel bomb**. *British equivalent:* **Anti-personnel bomb**.

Fragmentation grenade.—A hand grenade that shatters as it bursts, throwing small bits of metal in all directions.

Frangible grenade.—An incendiary grenade, consisting of a glass bottle filled with gasoline or a chemical, that ignites when the bottle is smashed.

Front.—*a.* The direction of the enemy.

b. The line of contact of two opposing forces.

c. The space occupied by an element measured from one flank to the opposite flank.

d. A Soviet combat formation that is equivalent to a **Group of armies**.

e. British equivalent: Same (in meanings *a*, *b*, and *c*.).

Frontage.—The space, in width, occupied or covered by a unit in any formation. *British equivalent:* Same.

Front line.—The line formed by the most advanced units, exclusive of local security, in any given situation. In a war of movement, the front line is the advanced line that is protected by artillery; in a war of position, the front line connects the advanced point in the defense system. Also called **line of battle**. *British equivalent:* Same.

Furlough.—An authorized vacation from military duty for an enlisted man. A **pass** is for three days or less; a furlough is for a longer period. A similar vacation authorized for an officer is called **leave**.

Fuze.—*a.* A mechanical or electric device used with a projectile, mine, bomb, or grenade to explode it at the desired time.

b. A train of powder used to detonate explosives or to fire pyrotechnics.

G-1, G-2, G-3, G-4.—*See* **General staff**.

Gabion.—A cylindrical basket woven with open ends, filled with earth, and used as a retaining wall in constructing field works. Sandbags are often used in place of gabions.

Gait.—Manner of movement of the horse, that is, the walk, trot, or gallop. *British equivalent:* Same.

Gait of march.—The gait at which the base of a mounted unit is moving at the instant considered. *British equivalent*: None.

Gallery.—*a.* Un underground passageway that is part of a military mining system.

    b. A compartment for storage of ammunition.

    c. An enclosed range for target practice.

Garand rifle.—A semiautomatic, gas-operated, clip-fed rifle. The Garand rifle has a caliber of .30 inch and weighs 8.56 pounds. It replaced the Springfield and Enfield bolt-action rifles as the standard equipment in the U.S. Army. The official designation is "U.S. Rifle, Caliber .30, M1."

Garrison.—*a.* A body of troops stationed at a military post.

    b. A military installation at which troops are stationed. In this meaning, also called *post*.

    c. To station troops at such a post or position.

Garrison cap.—A small olive-drab or khaki cap. Gold braid or black and gold braid on the cap of an officer, or colored cord braid on the cap of an enlisted man shows the arm or service of the officer or enlisted man. Formerly called a *field cap* or *overseas cap*.

Garrison ration.—A food allowance for one person prescribed (in peacetime) for all persons entitled to a ration. It is issued in the form of a money allowance.

Gas officer.—A staff officer in a unit smaller than a division assigned to advise on, and be responsible for, plans, equipment, and training for chemical warfare defense. In a division or higher unit, the staff officer having these duties is called the *chemical officer*.

Gas-operated.—To put in motion by the action of expanding gases. This term refers especially to the mechanism in certain types of automatic guns that are operated by deflecting a part of the escaping powder gases, so that its force operates the reloading mechanism through an opening in the barrel. The United States rifle, caliber .30, M1 is a gas-operated weapon.

General.—*a.* An officer in the United States Army who ranks next above a lieutenant general and next below a general of the Armies of the United States. A general is the equivalent to an admiral in the United States Navy.

    b. A title by which any general officer is addressed in conversation; an officer of one of the first four grades: general, lieutenant general, major general, brigadier general.

General attack.—*British term*: see *Assault, general.*

General classification test.—A test that is intended to measure the ability to learn quickly rather than the amount of knowledge a person has acquired. A general classification test includes tests on arithmetic and the meaning or words, and also simple problem questions. Enlisted men are given this test when they are inducted.

General counterpreparation.—A counterpreparation planned to meet a general attack and involving all the weapons capable of firing on the threatened front. *British equivalent:* **Defensive fire**.

General court-marital.—The highest type of court-martial, consisting of not fewer than five officers, having power to try any crime punishable by the Articles of War. **British equivalent**: Same.

General depot.—A large supply establishment for receiving, storing, and issuing supplies for more than one branch of the Army. A depot that serves a single branch is called a **branch depot**.

General dispensary.—An army medical establishment that provides medical and dental care for military personnel receiving treatment at the establishment but not staying there. A general dispensary is located in a large city, military district, or prescribed military area. An establishment at a military station is called the **station dispensary**.

General Headquarters.—The headquarters of the commander of all field forces. Often abbreviated as GHQ.

General headquarters reserve.—Troops of various arms and services not organically assigned to an army in the field, which are held for use as reinforcements or for separate missions under General Headquarters; GHQ reserves.

General hospital.—A hospital designed to serve general and special needs, equipped and staffed for special treatment of a professional nature. A general hospital affords better facilities than other hospitals for the study, observation, and treatment of serious, complicated, or obscure cases.

General map.—A small-scale military map showing a considerable area, used for general planning purposes.

General officer.—Any officer above the rank of colonel. Generals, lieutenant generals, major generals, and brigadier generals are general officers.

General of the Army.—An officer of the highest rank in the Army of the United States. He is above a general in rank. This rank is equivalent to a field marshal in most European armies. *British equivalent*: **Field marshal.**

General orders.—*a*. Permanent instructions issued in order form, that apply to all members of a command, as compared with **special orders**, which affect only individuals or small groups. Such orders are usually concerned with matters of policy administration.

b. A series of permanent guard orders that govern the duties of a sentry on duty. All soldiers are required to memorize them.

General outpost.—Stationary bodies of troops place at some distance from the main force, while at a halt, in camp or bivouac, or in battle position, to protect the main force from observation and surprise by the enemy, and to keep the enemy under observation.

General staff.—A group of officers in a division or larger unit who assist their commander in planning, coordinating, and executing operations. A general staff is usually divided into four sections: personnel (G-1), military intelligence (G-2), operations and training (G-3), supply and evacuation (G-4). In units smaller than the division, to include the battalion (or equivalent unit), these corresponding duties are assigned to officers designated as executive S-1, S-2, S-3, and S-4.

General Staff.—A body of officers detailed to the performance of staff duty in the War Department or with divisions and higher units. The general staff is headed by a Chief of Staff who may be assisted by one or more deputy chiefs. Each section is headed by an Assistant Chief of Staff. The sections of the General Staff are as follows: G-1, personnel; G-2, intelligence; G-3, operations and training; G-4, supply; and with the War Department, a fifth section, War Plans (which in wartime has become the Operations Division of the War Department, and is not to be confused with G-3). *British equivalent*: **The Staff (including the General Staff (or the "GS" or the "G") branch**, the **Adjutant-General's (or the "A") branch**, and the **Quarter-Master-General's (or the "Q") branch).**

General Staff Corps.—A branch of the Army made up of specially trained officers who are detailed to perform general duties either with the War Department or with divisions or larger units.

General staff with troops.—A part of the General Staff Corps including officers on duty with divisions or larger units, but not including those on duty with the War Department General Staff.

General supplies.—Supplies required for the maintenance of an organization other than ammunition or cleaning and preserving supplies. Forage, fuel, and food are general supplies.

General support.—Support provided by that artillery which supports the entire force of which it is a part. *British equivalent*: None.

Geneva Convention.—An agreement between European nations, in 1864, with later amendments, establishing rules for the treatment during war of the sick, the wounded, and prisoners of war.

Geographic code.—A system of code names for towns and other points, used in signal communication.

Geographic coordinates.—Latitude and longitude; north-south and east-west lines whose intersections are used to locate physical points on a map.

GHQ.—See *General Headquarters.*

GHQ reserve.—See *General headquarters reserve.*

Glider.—A heavier-than-air aircraft without a motor that is towed aloft or projected into the air by various means. It may glide to the ground or ride on upward-moving air currents. Gliders are also towed by airplanes.

Gooseberry.—A portable obstacle made of barbed-wire balls connected with a spiral of barbed wire. It is mainly used to block trenches.

Grade.—*a.* An indication of the quality or type of ammunition.

b. Rank in the Army. Grade applies to both officers and enlisted men, although usually applied to enlisted men and warrant officers.

c. The slope of a hill, road, or railroad.

Grand tactics.—Military operations conducted by large groups of troops.

Graphite.—One of the natural forms of soft carbon. It is used as a dry lubricant.

Grazing fire.—Fire whose path is close enough to the ground to hit a standing man for a considerable part of the range. Grazing fire is different from *plunging fire*, which is fire that strikes the ground at a high angle.

Grenade launcher.—An extension attached to the muzzle of a rifle or carbine that converts the gun into a device for firing rifle grenades.

Grid.—See *Military grid.*

Grid coordinates.—A method of locating a point in a north-south direction or an east-west direction in reference to the grid lines on a military map.

Grid line.—One of the lines in a grid system; a line used to divide a map into squares. East-west lines in a grid system are X-lines, and north-south lines are Y-lines.

Grid north.—The direction in which the north-south grid lines on a military map point, generally not the same as true north; Y-north. Military maps ordinarily indicate true north, magnetic north, and grid north in order to give complete orientation in direction.

Groove.—a. Any one of several spiral channels cut in the bore of a gun to rotate the projectile when it is fired. The raised surfaces between the grooves are lands; rifling consists of both lands and grooves.

b. A small, narrow channel cut around a projectile near its base, used to hold grease in heavy ammunition and to fasten the cartridge case in fixed ammunition.

Ground alert.—A status in which an aircraft on the ground is fully serviced, armed and with combat crews in readiness to take off 15 minutes after receiving orders to perform a mission.

Ground alert method.—One of three methods of using fighter aviation in air defense. With this method the defending fighter force is held ready on the ground for immediate take-off upon receiving warning of approaching aircraft. Other methods of using fighter aviation are the *air alert method* and the *search patrol method*.

Ground Forces.—See *Army Ground Forces*.

Ground observation.—Observation of enemy or friendly positions, activities, fire, etc., from a point on the ground. terrestrial observation.

Ground readiness.—A status wherein an aircraft can be serviced and personnel alerted to leave the ground for a mission within 2 hours.

Ground strips.—*British term*: see *Identification panels*.

Group.—a. A tactical unit of the harbor defense formed for the purpose of fire direction.

b. A U.S. Air Air Forces unit composed of two or more squadrons of a single class of aviation. Groups usually contain only one type of aviation, although composite groups may be formed where the situation demands. *British equivalent:* **Wing**.

c. *British equivalent:* A formation of the R.A.F. composed of two or more wings. (*Note that the organizational hierarchy of R.A.F. is Group, Wing, Squadron, compared to the U.S.A.A.F hierarchy of Wing, Group, Squadron.*)

d. An armored force unit composed of two or more GHQ reserve tank battalions either or both of light, medium, or heavy tanks. See also **Combat team** and **Groupment.**

  e. British term: A body of troops, based on a formation, unit, or sub-unit specified, with the addition of other arms under command as required for a specific operation, e.g., brigade, regimental, company group.

  f. British term: Under the group system, the main body or fighting portion of each unit was divided into groups for deployment and receiving orders. The four main groups concerned in deployment were:

> **R (recce) group.**—Contains all the personnel a commander may require to help him during his reconnaissance, and in preparation of his plan.
>
> **O (order) group.**—Contains the minimum personnel necessary to receive a commander's orders, so as to put his plan into operation.
>
> **F (fighting) group.**—Contains the personnel, vehicles, and equipment necessary to carry out the commander's plan.
>
> **T (transport) group.**—Contains the personnel, equipment, and A echelon transport not required in F group.

Grouping.—*British term:* see **Groupment.**

Groupment.—*a.* A temporary field artillery grouping of two or more battalions or larger units which have the same tactical mission. *British equivalent:* **Grouping**

  b. A tactical command in the coast artillery corps. containing two or more groups or separate batteries whose fields of fire cover a certain water area, together with personnel and matériel required for its employment as a unit. *British equivalent:* **Group.**

Group of armies.—A tactical unit consisting of two or more armies with suitable reinforcements placed under a designated commander for the accomplishment of a particular task, the execution of which requires coordination and control by one commander; army group. The group (task force) may operate under the War Department or under a theater commander.

Guard.—*a.* An individual or group set the task of protecting an encampment, station, or the like, from a surprise attack, or to prevent the escape of hostile action of prisoners. A guard keeps a systematic watch of the place or persons entrusted to his care. An interior guard keeps watch within the limits of a post, camp, or station.

  b. To act as a guard.

  c. A curved piece of metal on a gun within which the trigger is located and which protects the trigger. In this meaning, usually called **trigger guard**.

  d. A position in bayonet practice in which one holds a rifle with the bayonet pointed at the enemy's throat, ready for instant attack.

"Guards."—A Soviet title appended to any unit which distinguishes itself in combat.

Guerrilla (or partisan) warfare.—Irregular war carried on by independent bands. *British equivalent:* Same.

Guide.—An individual who leads or guides a unit or vehicle over a predetermined route or into a selected area.

Guidon.—*a*. A flag, streamer, or pennant carried by Army units for identification. It is usually cut swallow-tailed.

  b. A soldier who carries the guidon.

Gun.—Any firearm; especially, a cannon that has a relatively long barrel, low angle of fire, and high muzzle velocity. Guns are classified according to their method of loading, angle of fire, or tactical use.

Guncotton.—An explosive made by treating cotton with nitric and sulphuric acids; nitrocotton. Guncotton is used to carry the flame to the burster in some projectiles, and in electric priming devices. Guncotton is also used in making certain high-grade smokeless powders. It is a **nitrocellulose** with a high nitrogen content.

Gunpowder.—See **Black powder**.

Gun section.—*a*. In the field artillery, a subdivision of a battery consisting of a gun and caisson with proper personnel and necessary equipment.

  b. In the coast artillery, one gun in its emplacement, with assigned personnel and necessary equipment.

  c. In the armored command, one of the three sections constituting the reconnaissance platoon of an armored reconnaissance company. A gun section consists of a self-propelled assault gun and a half-track ammunition carrier.

Gun fire targets.—*British term:* see **Fire on targets of opportunity**.

Gyro control.—*a*. The gyroscopic unit of an automatic pilot.

  b. A gyroscopic device for maintaining the steady position of a structural unit, such as a device that steadies a gun in a moving and pitching tank or the device that reduces the pitch and roll of a ship. In this meaning, also called **gyro-stabilizer**.

Gyro-stabilizer.—A gyroscopic device for maintaining the steady position of a structural unit, such as a device that steadies a gun in a moving and pitching tank or the device that reduces the pitch and roll of a ship. Also called *gyro control*.

Half-track vehicle.—A combination wheeled and track-laying vehicle, usually steered by an ordinary front-wheel arrangement, and propelled from the tractor-track arrangement behind the front wheels; half-track.

Hand arms.—All weapons such as submachine guns, automatic pistols, revolvers, or swords, that can be carried and used by hand; hand weapons.

Hand grenade.—A small bomb, thrown with the hand, that explodes on impact or by time fuze. Hand grenades are divided into three general types: fragmentation, those containing a chemical filler, and those designed for training or practice.

Handset.—A telephone in which the transmitter, receiver, and connecting handle form a single piece.

Hand weapon.—See *Hand arms*.

Hangfire.—A temporary failure or delay in the action of a primer, igniter, or propelling charge. For a few seconds it cannot be distinguished from a complete failure, or *misfire*.

Harass.—To annoy and disturb the enemy by fire, raids, frequent small attacks, etc. *British equivalent*: Same.

Harassing agent.—A chemical agent used to force masking and thus slow up the enemy operations. A harassing agent produces irritating effects only, while a *casualty agent* is powerful enough to cause death. *British equivalent*: Same.

Harassing fire.—Fire delivered to interfere with and annoy the enemy, to keep his troops alerted unnecessarily, and to lower his efficiency and morale. See also *Interdiction fire*. *British equivalent*: Same.

Harassing tactics.—Tactics intended to annoy and hinder rather than to destroy, as in *"hit and run"* tactics.

Harbor.—A sheltered body of water of sufficient depth to enable a ship to find shelter in it from the storms of the high seas. *British equivalent*: *Harbour*.

Harbor defense.—A highly organized administrative and tactical Army command established to defend a limited portion of a coastal area primarily against attacks from the sea. *British equivalent*: *Harbour defence* (the command may be made up of naval, military, and air forces, and commanded by an officer specially appointed).

Harbour .—*British definition*: *a*. A sheltered body of water of sufficient depth to enable a ship to find shelter in it from the storms of the high seas.

  b. A lying-up area for armored formations or unit.

  c. An area under cover from the air, which is both a position of readiness and a place where maintenance and refilling operations are carried out.

Harbouring party.—*British term*: see **Quartering party.**

Hasty field fortification.—A fox hole, trench, gun emplacement, or obstacle constructed under fire or under threat of immediate attack. A hasty field fortification is generally less elaborate than a deliberate field fortification, which is constructed more carefully and not under fire or threat of attack.

–head.—*British definition*: A suffix to terms denoting communications (e.g., railhead, riverhead, etc.) forming part of the organization of the **Lines of Communication.** The locality so named denotes the point of transhipment from one form of **communication** to another, and is included in the Lines of Communication.

Haversack.—The principal part of the field pack of a soldier, consisting of a canvas case in which the field equipment of the soldier is packed; knapsack. The haversack has shoulder straps and harness to support the pack on the back.

Headquarters.—*a*. A place from which the chief or commanding officer of an organization issues orders and conducts administrative and tactical work. *British equivalent:* Same.

  b. A group of officers and enlisted men forming the directing and coordinating head of a tactical unit, usually including the unit commander, his staff, and attached personnel, that performs duties in relation to administration, intelligence, communications, and other necessary activities. In this meaning, also called **command element.** *British equivalent:* Same.

Head of column.—First element of a column in order of march. *British equivalent:* Same.

Head of service.—*British term*: The senior officer of a service in a theater of operations.

Heavy artillery.—*a*. Artillery pieces of largest caliber, usually 155 millimeters or larger. Sometimes the 105-millimeter howitzer is classed as heavy artillery. Other classes are light artillery and medium artillery.

  b. Artillery units that use such pieces.

Heavy coast defence guns.—*British term*: **Seacoast artillery (Coast Artillery Corps)**.

Heavy machine gun.—A classification of machine guns including the .30-caliber water-cooled machine gun, and all .50-caliber machine guns. It is often used, however, to refer to the .30-caliber water-cooled machine gun, specifically, in order to distinguish it from other machine guns of the same caliber. *British equivalent*: **Medium machine gun**.

Heavy tank.—A tank of over 40 tons in weight. Heavy tanks also carry the greatest armament and armor. Tanks are usually classified as light (up to 25 tons), medium (25 to 40 tons), and heavy (over 40 tons).

Heavy weapons.—All weapons, such as mortars, howitzers, guns, and heavy machine guns, which are usually part of infantry equipment.

Hedgehog.—A portable obstacle made of three poles or logs crossed and bound at their centers to make a framework of hour-glass shape, the whole laced with barbed wire.

H-hour.—The hour set for an attack or other operation to begin. Also called **zero hour**. *British equivalent*: Same.

High-angle fire.—Fire delivered at elevations greater than the elevation of maximum range; fire the range of which decreases as the angle of elevation is increased. Mortars deliver high-angle fire.

High explosive.—Any explosive that goes off or burns so rapidly that it produces a shattering effect; detonating explosive. A high explosive is therefore suitable as a bursting charge in bombs and projectiles, while a low explosive is suitable for use in propelling charges in guns, or for ordinary blasting. It is commonly called HE.

Hit.—*a.* Impact on a target.
  b. To strike a target.

"Hit and run" tactics.—Repeated attacks and withdrawals, in which the attacker refuses to stay and fight out an engagement. "Hit and run" tactics are intended to harass the enemy.

Hold (verb).—To retain physical possession. *British equivalent*: Same.

Holding and reconsignment point.—A rail or motor center with considerable capacity to which cars or trucks may be sent and at which they may be held until their destination becomes known or until the proper time for them to be moved farther toward their destination. *British equivalent*: **Railway siding** or **motor park**.

Holding attack (or secondary attack).—The part of the attack designed to hold the enemy in position and prevent the redistribution of his reserves; it is ordinarily directed toward the hostile front. *British equivalent*: Same.

Holding force.—A force assigned to holding a place or position; a force that carries out a holding attack; holding element.

Home Forces.—*British term*: All field forces, including the Home Guard, located in the United Kingdom responsible for the defense of the British Isles. It consists of all corps, divisions, and separate units of the British Army assigned to defend against invasion.

Home Guard.—*British term*: Originally organized hastily in May 1940 as the **Local Defence Volunteers**, it consists of volunteer unpaid, part-time troops formed into units for local defense of communities, airfields, and communications, and for general observation purposes.

Horse artillery.—Light mobile artillery in which the carriages are drawn by horses or mules and the cannoneers are mounted on horses.

Horse-drawn artillery.—Light or medium artillery moved by teams of draft horses.

Horse length.—A term of measurement. For convenience in estimating space, a horse length is considered 3 yards; actually, it is about 8 feet. *British definition*: 8 feet.

Howitzer.—An artillery piece with a medium-length barrel, between that of a mortar and a gun in length, operating with a high angle of fire (as high as 65 degrees of elevation) and using a medium muzzle velocity. The high angle of fire allows a howitzer to reach targets hidden from flat trajectory guns.

Hull defilade.—A position taken by a combat vehicle whereby only the fighting compartment (turret) is exposed to hostile fire or view; hull down. See also **Turret defilade**. *British*: **Hull down**.

Hull down.—See **Hull defilade**.

Hutment.—*British term*: see **Cantonment**.

Incendiary.—*a.* Any chemical agent that produces enough heat to set anything on fire.

   b. An aerial bomb which, upon striking, releases a chemical agent that starts a fire.

   c. Used to start a fire, such as an incendiary bomb. Oil and thermite are common **incendiary agents**. An **incendiary grenade** bursts upon striking and releases chemical agents that start fires. There is also *incendiary ammunition,* an incendiary projectile, etc.

Identification.—*a.* A special mark by which personnel, organizations, or equipment are identified.

 b. Identifying personnel, equipment, etc., from distinctive markings or other means.

 c. The process of determining by methods other than visual whether an object is friendly or not. The process of determining whether an object is friendly or not by visual methods is called *recognition.*

Identification mark.—Distinctive markings, especially on an aircraft or vehicle, that identifies its nationality, type, manufacturer, etc.

Identification panels.—Panels of cloth or other easily handled material which are displayed by ground troops to indicate to friendly aircraft the position of a unit. The panels are usually laid out in a group of three numbers, called an *identification group.* British equivalent: *Ground strips.*

Identifications.—Any distinctive marks or other means by which personnel, organizations, or equipment are identified. *British equivalent:* Same or *(distinguishing) marks* or *markings* or *"splashes."*

Intrench.—To dig any kind of trenches; to fortify with trenches or field fortifications; entrench.

Intrenchment.—A fortification consisting of a trench and a bank; entrenchment.

Inverted wedge formation.—A formation resembling an inverted triangle, in which two units advance abreast of each other and a third unit follows in the rear.

Immediate message.—*British term*: see *Urgent message.*

Immediate counter-attack.—*British term*: see *Counterattack.*

Immobilize.—To tie down. To deprive of mobility. *British equivalent:* Same.

Impedimenta.—Military personnel and supplies and equipment that are taken into the field with the troops, but not into actual combat. These articles are called impedimenta only when they accompany troops in movement.

Impress.—*a.* To seize by force for public use.

 b. To force men to serve in the armed forces. Men are impressed in an emergency to help defend the country.

Imprest.—*British term*: An advance of public money for expenditure on the public service, e.g. payment of troops, local purchases, etc.

Important message.—*British term*: see **Priority message.**

Incendiary agent.—An agent used primarily for setting fire to material. *British equivalent*: Same.

Indent.—*British term*: see **Requisition.**

Indirect fire.—Gunfire delivered at a target that cannot be seen from the gun position; gunfire in which the gun is aimed for direction and elevation by means other than sighting at the target.

Indirect laying.—Laying in which the line of sighting is directed upon a fixed object other than the target. *British equivalent*: Same.

Indirect support.—Help given by aircraft to a combat unit. Indirect support includes cutting supply and communication lines in the enemy's rear and destroying enemy airplanes; **direct support** includes observation, transmission of messages, and other particular tasks.

Individual equipment.—Those supplies necessary to enable the individual to function as a soldier. *British equivalent*: **Personal clothing and equipment.**

Individual reserves.—Those supplies carried on the soldier, animal, or vehicle for his or its individual use in an emergency.

Infantry.—Troops trained, equipped, and organized to fight on foot.

Infantry tank.—*British term*: Tanks designed to support infantry attacks on field fortifications, preceding the infantry and attempting to overrun the enemy defenses; army tanks. They are slower and more heavily armored than **cruiser tanks**. The role of British infantry tanks is one which was not normally envisaged by U.S. armored tactics, although in unusual circumstances U.S. heavy tanks might be called upon to perform a similar mission.

Infiltrate.—To pass troops in relatively small numbers through gaps in the enemy position or his field of fire. For example, to advance individuals by bounds during an attack. *British equivalent*: Same.

Information center.—A station in an aircraft warning service where data about enemy aircraft are gathered and plotted on a map, and are then sent to antiaircraft defense units.

Initial aiming point.—A point on which a gun is sighted to establish a reference line from which direction angles for targets are measured. From this reference line, other aiming points that give the direction of the target are measured off. This method of aiming is used in *indirect laying.*

Initial assembly area.—A place where troops are originally concentrated in preparation for action. When the troops have been organized in the initial assembly areas they move up to a final assembly area, from which they move into combat.

Initial firing position.—The first position from which attacking troops open fire on the enemy.

Initial point.—A point at which a moving column is formed by the successive arrival of the various subdivisions of the column. *British equivalent: Starting point.*

Initial requirements.—Those supplies required to meet the original demands incident to field operations.

Inner harbor area.—The entire water area of a fortified harbor inside the inner entrance of all the entrance channels to the harbor.

In position.—A term which indicates that the weapons of a unit are in position and ready to fire and that necessary systems of observation and communication have been established. *British equivalent:* Same.

In readiness.—A term which indicates that an artillery unit is held near one or more possible positions, prepared to move quickly into position when ordered. This term may be applied to other units to indicate a state or condition of preparedness. *British equivalent:* Same.

Inshore patrol.—A part of the naval local defense forces operating generally within a defensive coastal area and controlling shipping within a defensive sea area.

Inspector general.—An officer of the Inspector General's Department who examines and reports on property, records, accounts, and all other matters that affect the efficiency and economy of the Army. The special staff of a division or higher unit usually includes an inspector general, who advises the commander on matters affecting the efficiency and economy of the command.

Installation.—*a.* A military organization in a fixed place, together with its buildings and equipment. A depot, hospital, post, or station is an installation.

b. British: A locality organize for a specific purpose of a service.

Insubordination.—Disobedience; resistance to authority; refusal to obey lawful orders; disrespect.

"In support."—*British term:* see *Direct support.*

Integrity of tactical units.—The maintenance of complete tactical units.

Intelligence.—*a.* The work of the intelligence personnel of a military organization in gathering, evaluating, and disseminating information of military value.

    b. Information of military value, gathered, evaluated, and disseminated by the intelligence personnel.

    c. A division or section of a military unit that gathers, evaluates, interprets, and disseminates information of military value. In all meanings, also called *military intelligence.*

Interception.—Engaging an enemy force in an attempt to hinder or prevent it from carrying out its mission. Interception is usually carried out by airplanes.

Interceptor.—A type of fighter plane with a high rate of climb and speed, used to intercept enemy aircraft.

Intercept station.—A radio station that copies enemy radio traffic for the purpose of obtaining information, or friendly traffic for the purpose of supervision.

Intercommunication.—*British term*: The means of transmission of all orders and information, by which the close cooperation of all forces in the field is ensured. The means include the service provided by the Royal Corps of Signals, by regimental signalers and orderlies, by liaison officers, and by the postal service.

Interdict.—To prevent or hinder the use of an area or route by the application of chemicals or fire, or both. *British equivalent*: None.

Interdiction fire.—Fire delivered on certain areas or routes to prevent or hinder their use. For example, fire which seeks to make untenable certain areas or to interrupt movement over certain routes of communication (may also be affected in certain situations by contamination of the ground by persistent gas or by land mines). Interdiction may be partial or complete.

Intermediate depot.—A general or branch depot located in the intermediate section of the communications zone in a theater of operations..

Intermediate objective.—The objective whose attainment precedes and is usually essential to the attainment of the final objective. *British equivalent*: Same.

Intermediate position.—A position between the assembly positions and the line of departure which may be occupied temporarily by units of the attacking echelon for coordination of the attack, for further reconnaissance, or last-minute servicing of vehicles. (*Also British term: see* **Delaying position**)

Intermediate scale maps.—Maps normally of a scale from 1:200,000 to 1:500,000, intended for planning strategic operations, including the movement, concentration, and supply of troops. *British equivalent:* **Small-scale map** (includes U.S. definition of both small-scale and intermediate-scale maps).

Intermediate section.—That portion of the communication zone lying between the advance and base sections. *British equivalent:* None.

Interpretation of information.—An analysis of information to determine its probable significance in the existing situation.

Interval.—Space between individuals or elements of the same line. Between mounted troops, it is measured from knee to knee. Between dismounted troopers, it is measured from elbow to elbow. Between vehicles, it is measured from hub to hub. *See also* **Time interval.** *British equivalent:* Same.

Inverted wedge formation.—A formation resembling an inverted triangle, in which two units advance abreast of each other and a third unit follows in the rear.

Irritant smoke.—A chemical agent which causes sneezing, coughing, lacrimation, or headache followed by nausea and temporary physical disability. *British equivalent:* **Toxic smoke.**

Island of resistance.—A small group that still resists the enemy although it is cut off from its own forces.

Isolation.—A complete cutting off from other forces, nations, etc.

Issue.—A delivery of supplies. Specifically, the delivery of supplies of any kind by a supply department to responsible persons authorized to receive them on behalf of their organizations. The supplies so delivered. To send out officially or publicly, as orders or communiques. To emerge or sally forth, as from a defile or fortress. *British equivalent:* Same.

Joint operations.—*See* **Combined operations.**

Joint plan.—A war plan whose purpose is to establish the basis and prepare the necessary plans for joint action by the Army and Navy in a given situation. *British equivalent:* Same.

Journal.—A chronological record of events affecting a unit or staff section. *British equivalent:* **War diary.**

Judge advocate.—*a.* A staff officer who acts as the chief adviser on military law to a commander who has the authority to appoint courts-martial. In this meaning, usually called **staff *judge advocate***.

   b. An officer designated by the appointing officer to prosecute an action in the name of the Government, in a general or special court-martial. In this meaning, usually called ***trial judge advocate***.

Jump area.—A locality assigned by a commanding officer for the landing of parachute troops in a combat operation. A jump area is usually behind enemy lines. Also called ***landing area***. *British equivalent:* ***Dropping zone***.

Jumpmaster.—An officer or noncommissioned officer who controls the jumping of parachute troops and the dropping of their equipment from aircraft.

Jump off (verb).—To leave one's lines for an attack.

Jump-off (noun).—A start of a planned ground attack.

Jump-off line.—A line from which an attack starts. Also called ***line of departure***.

Jump-off point.—A point at which parachute troops must jump from the aircraft in order to land in the area chosen for their operation.

Junction point.—*British term:* A distinctive topographical feature, on or in a neighborhood of a unit or formation boundary, generally selected from the map before an operation to facilitate coordination between the flanks of adjacent formations or units during that operation. See ***Limiting point***.

Junior officer.—An officer in the Army below field grade. A captain, a first or a second lieutenant is a junior officer.

Key point.—A tactical locality affording observation and communication, the possession of which is vital to the success of an engagement. *British equivalent:* Same or ***Vital point***.

Key terrain.—A part of an area that gives an advantage in combat to the side holding it.

Khaki.—A dull yellowish-brown color. Uniforms in this color are often called khakis.

King's Regulations.—*British term:* see ***Army regulations***.

Kit.—*a.* Equipment and personal necessities that a soldier carries with him.

   b. An outfit of tool, instruments, chemical detecting apparatus, etc.

c. A small bag or case for carrying such equipment or such outfit.

K ration.—One of the emergency field rations used when other rations are not available.

Lacrimator.—A chemical agent which causes a continuous flow of tears and intense, though temporary, eye pains. Also called ***tear gas***. *British equivalent*: ***Lachrymator***.

Ladder.—Bursts of three rounds fired rapidly by a single gun pointing in the same direction. The bursts are at intervals of 300 yards in range. The first round is fired at the greatest range.

Landing area.—*a.* A level field suitable for the landing and take-off of aircraft.

b. A locality assigned by a commanding officer for the landing of parachute troops in a combat operation. A landing area usually is located behind enemy lines. In this meaning, also called ***jump area***. *British equivalent*: ***Dropping zone***.

Landing attack.—An attack against a defended shore by troops who come ashore from boats, rafts, barges, or amphibian vehicles.

Landing barge.—A special type of flat-bottomed boat used to carry troops and combat equipment ashore for a landing attack.

Landing boat.—A boat specifically designed to carry troops and combat equipment ashore for a landing attack. It may be of any size, varying from small surf craft with a capacity of one squad to large self-propelled barges capable of landing tanks.

Landing craft.—Any vessel used to carry men, equipment, and supplies ashore.

Landing field.—A field or system of runways suitable for the landing and take-off of airplanes. *British equivalent*: Same.

Landing-places.—*British term*: In combined operations, the actual places selected for breaching craft or building piers.

Land mine.—A container filled with explosives or chemicals, placed in the ground or lightly covered. It is usually set off by the weight of vehicles or troops passing over it.

Lanyard.—*a.* A strong cord with a small hook at one end, used to fire certain kinds of cannon, pyrotechnic projectors, etc.

b. A strong cord worn around the neck or shoulder and attached to a side arm to keep it from being lost.

Large scale maps.—Maps normally of a scale not greater than 1:20,000 intended for the technical and tactical needs of the combat arms. *British equivalent*: Same.

Large units.—Divisions and larger units.

Lateral.—An underground gallery that is constructed parallel to the front line, and from which other parallel galleries for attack, defense, and listening are projected forward toward the enemy. A lateral differs from a **fishbone**, which is a series of independent galleries cut in the direction of the enemy.

Lateral action.—Action toward the side. A bridgehead may be enlarged by lateral action along the river.

Latrine.—A privy or toilet in a bivouac or camp; washroom and toilet facilities in barracks. *British equivalent*: Same

Laying.—*a*. The process of pointing a gun for a given range, or for a given direction.

    b. Assembling a mosaic by pasting aerial photographs to a mount.

Lay-on-me method.—A method of aiming in which a battery observes a friendly airplane that flies at a specific altitude along a gun-target line and makes a sudden turn when directly over the target. Firing data is calculated from the direction, altitude, and distance of the airplane at the moment it makes its turn.

Lead.—*a*. A distance ahead of a moving target that a gun must be aimed in order to hit the target.

    b. The vertical and lateral angles between the gun-target line and the axis of the bore at the moment of firing at a moving target.

    c. To aim a gun ahead of a moving target.

    d. One target length, as it appears to the gunner used for measuring lead.

    e. The spacing between the heads of successive vehicles, march units, columns, etc., as they to forward in a line.

Leading.—The acts of a commander in controlling his unit by personal direction; the term used to designate a method of marching whereby the trooper dismounts but continues the march, leading his mount.

Leading fire.—Fire delivered ahead of a moving target to allow for its motion.

Leading troops in attack.—*British term*: see **Attacking echelon**.

Leaguer.—*British term*: see **Bivouac**.

Leapfrog.—To advance the elements of a command in the attack by passing successively through or by the other elements. *See also* **Passage of lines**. *British equivalent*: Same.

Leapfrogging.—*British term*: see **Passage of lines**.

Leave.—Permission for an officer to be absent from duty. A *furlough* is permission for an enlisted man to be absent from duty. Leave is also called *leave of absence*.

Lee-Enfield rifle.—The British standard-issue rifle, caliber .303. It is a bolt-type, breech-loading magazine rifle. The No. 4, Mk. I replaced the No. 1, Mk. III. Essentially, both types were like the U.S. rifle, caliber .30, M1903 (Springfield).

Leggings.—An extra outer covering of cloth or leather for the legs between the knee and foot.

Letter(s) of instruction.—A means by which the plans of superior commanders are communicated and which regulate movements and operations over large areas and for considerable periods of time. See also *Combat orders* and *fragmentary orders*. *British equivalent*: *Operation instruction(s)*.

Letter order.—An official military order or instructions sent in the form of a letter to the person or persons concerned.

Lewis gun.—A British medium (heavy) machine gun, caliber .303. A "heavy" machine gun by U.S. terminology, it was classified as a "medium" machine gun by the British. Air-cooled, gas-operated. Although obsolete, it was still in use by the Home Guard, but not in the field except occasionally as a weapon for ground defense. In the Home Guard it was used extensively for beach defense, and could be mounted on an antiaircraft mount for use against low-flying airplanes.

Lewisite.—A chemical agent that appears in the form of a dark brown, oily liquid or as a colorless gas, with a geranium-like odor. Its arsenic content, in either the gas or liquid form, causes extreme injury to skin and lungs.

Liaison.—The connection established between units or elements by a representative—usually an officer—of one unit who visits or remains with another unit. *British equivalent*: Same.

Liaison agent.—A person who maintains contact between a command post and other headquarters in the field, or between the headquarters of different units, when exchange of information is necessary.

Liaison airplane.—An airplane used as a courier to maintain contact between various parts of a military force. Liaison airplanes are usually slow and do not have armor or armament. They are able to operate from small areas.

Liaison officer.—An officer from the headquarters of one unit sent to represent his commander at the headquarters of another unit and to keep the two units in close touch. *British equivalent:* Same (noncommissioned officers may carry out this duty within units).

Lieutenant.—*a.* A company officer in the Army who ranks next below a captain and next above a warrant officer. There are two grades, first and second, the first being the higher rank. A lieutenant is usually in command of a platoon, or of an aircraft and its combat crew.

b. An officer in the Navy who ranks next above a lieutenant (junior grade) and next below a lieutenant commander. He has a rank equivalent to that of a captain in the Army.

Lieutenant colonel.—An officer in the Army who ranks next below a colonel and next above a major. A lieutenant colonel is usually in command of a battalion or a squadron. He may also be in charge of any part of a regiment larger than a battalion, or a regiment in absence of its commanding officer, or of any part of a group larger than a squadron, or a group in the absence of its commanding officer.

Lieutenant commander.—An officer in the Navy who ranks next below a commander and next above a lieutenant. A lieutenant commander has a rank equivalent to that of a major in the Army.

Lieutenant general.—An officer in the Army who ranks next below a general and next above a major general. A lieutenant general is usually in command of a corps or an air force.

Lieutenant (junior grade).—An officer in the Navy who ranks next below a lieutenant and next above an ensign. A lieutenant (junior grade) has a rank equivalent to that of a captain in the Army.

Light artillery.—*a.* Artillery pieces up to the size of, and usually including, the 105-millimeter howitzer. Other classes are **heavy artillery** and **medium artillery.**

b. Artillery units that use such pieces.

Light bombardment airplane.—A relatively small-sized bombing plane; light bomber. It carries a smaller load of bombs and has a shorter operating radius than a medium or heavy bomber, but it is more maneuverable and can operate at lower altitudes than either of the heavier types.

Lighter.—A boat or flat-bottomed barge used for loading and unloading ships.

Lighter-than-air aircraft.—Balloons and airships; aircraft supported by he lift of a gas that is lighter than air. *British equivalent:* Same.

Light machine gun.—Any machine gun including, and lighter than, a .30-caliber, air-cooled machine gun.

Light tank.—A tank of 25 tons or less. Light tanks are speedy and easily maneuvered, and carry light armor and armament. Tanks are usually classified as light (up to 25 tons), medium, (25 to 40 tons), and heavy (over 40 tons).

Limber.—*a.* A two-wheeled vehicle to which a gun or caisson is attached for transport.

    b. To attach a gun or caisson to its limber.

Limited attack.—An attack restricted to a single, set objective.

Limiting points.—Designated points where the several lines in a defensive position or outpost shall cross the unit sector boundaries, used to insure coordination between adjacent units. *British equivalent:* **Junction points.**

Limit of fire.—A boundary marking off the area on which gunfire can be delivered.

Line.—A formation in which the next lower subdivisions of a command are abreast of one another. *British equivalent:* Same.

Line ahead.—*British term:* see **Column.**

Linear defense.—A method of defense in which the defensive units and weapons are spread along a defense line in breadth rather than in depth.

Line maintenance.—Minor maintenance consisting of inspection, servicing, and repair of vehicles, aircraft, etc. It is performed in the field by the maintenance section of the crew of a vehicle or plane.

Line of battle.—A line formed by the most advanced unit in any given tactical situation. Also called *front line.*

Line of communications area.—*British term:* The area between the base ports and the rear boundaries of armies, through which the various means of transportation run. This may be divided into two or more L of C areas, which are subdivided into L of C. sub-areas. See also: **Communications zone.**

Line of communication terminal.—*British term:* A general term to cover road, rail, sea, or airhead, used when no specific head is referred to.

Line of defense.—Any position taken by combat forces for the purpose of defending an area or objective.

Line of departure.—A line designated to coordinate the departure of attack elements. Also called **jump-off line**. *British equivalent*: **Start(ing) line**.

Line of drift.—A natural route along which wounded men may be expected to go back for medical aid from a combat position.

(Line of) forward (foremost) defended localities.—*British term*: see **Main line of resistance**.

Line officer.—An officer belonging to a combatant branch of the Army; officer of the line.

Line of observation.—The line occupied by the observation elements of the outpost position. The line from a position finder to a target at the instant of a recorded observation. *British equivalent*: Same.

Line of operations.—A line along which a combat force operates in carrying out its missions. It begins at the base of operations and extends in the direction of movement towards the enemy.

Line of outposts.—*British term*: see **Outpost area**.

Line of position.—*a*. A straight line extending from a gun or a position-finding instrument to a point, especially a target. In this meaning, also called **line of site**.

b. A line extending from an observer to a point of known location. Tow such lines crossing each other at the observer's position determine the point of location in navigation or survey.

Line of resistance.—A line along which an enemy attack is met by organize resistance.

Line of retirement.—*a*. The direction of retirement of a retreating force to the rear. In this meaning, also called **line of retreat**.

b. A line at which scouting aircraft turn and start to search to the rear.

Line of retreat.—The direction of retirement of a retreating force to the rear. Also called **line of retirement**.

Line of retirement.—A direction of retirement of a retreating force to the rear. Also called a **line of retreat**.

Line of sight.—*a*. A line of vision; optical axis of a telescope or other observation instrument.

b. A straight line connecting the observer with the aiming point; the line along which the sights are set.

Line of site.—A straight line extending from a gun or position-finding instrument to a point, especially a target. Usually called a line of position.

Line of skirmishers.—A line of mounted or dismounted men in staggered formation at extended intervals.

Line of supply.—A network of railways, waterways, roads, and air routes along which supplies and equipment are moved from bases or supply points in the rear to tactical units farther forward.

Line of the Army.—All the combatant branches of the Army. These include the Coast Artillery Corps, Field Artillery Corps, Army Air Forces, Cavalry, and Infantry.

Line of withdrawal.—The direction of retirement of a force during a withdrawal.

Line overlap.—*British term*: see **Reconnaissance strip**.

Lines of action.—The possible plans open to a commander in a particular situation. *British equivalent*: **Courses of action** or **courses open**.

Lines of communication.—*a.* The network of railways, waterways, and roads which lead into the combat zone from administrative establishments located in the communications zone or in the zone of the interior.

b. The system of communications in a theater of operations between the bases inclusive and the rear limit of administration by formation commanders, along which the requirements of the field army move.

c. British: The system of communication to a theater of operations and in a theater of operations. The latter starts at at the base ports and ends at the terminal of each means of communication, i.e., rail, motor transport, inland water transport. See also: **Supply lines**.

Listening post.—A concealed or sheltered position established in advance of a defensive line for early detection of the enemy's movements. *See also* **Cossack post**. *British equivalent*: Same.

Litter.—A stretcher for carrying sick, wounded, or dead persons; stretcher.

Litter bearer.—A person who helps carry a stretcher. Litter bearers are usually medical aid men who carry wounded back to battalion aid stations fro emergency treatment.

Litter carrier.—A light two-wheeled cart operated by two men and used for carrying a stretcher with a sick, wounded, or dead person.

Litter relay point.—A point where litter bearer squads change the mode of transport, such as from hand litter to wheeled litter, or where a new litter bearer squad takes over further movement of the patient. *British equivalent*: None.

Livens projector.—A mortar installed in the ground to project chemical agents. *British equivalent*: Same.

Loyd carrier.—*British term*: A type of British armored carrier.

Loading point.—Any location at which trains, trucks, aircraft, etc., are loaded with personnel, supplies, and equipment.

Local assault.—See **Assault, local.**

Local attack.—*British term*: see **Assault, local.**

Local counterattack.—A counterattack made by local reserves to retake ground or to force the enemy to use up his own reserves.

Local counterpreparation.—A counterpreparation covering only that portion of the front threatened by a local attack and normally employing only the division artillery supporting the threatened front. *British equivalent*: **Defensive fire.**

Local Defence Volunteers.—*British term*: see **Home Guard.**

Local sector.—One of the subdivisions of a subsector.

Local security.—A security element, independent of any outpost, established by a subordinate commander to protect his unit against surprise and to insure its readiness for action.

Locations for sector controls.—*British term*: see **Critical points.**

Logistics.—That branch of military art that comprises everything relating to movement, supply, and evacuation. *British equivalent*: **Transport, supply, and quartering of troops.**

Lorry.—*British term*: The British Army made a distinction between trucks and lorries, "truck" being used for a load-carrying vehicle of 1 long ton of less, and "lorry" for a load-carrying vehicle of 30-cwt (1.5 long tons) or more. In addition, the term "van" was used for a truck with a fixed top, and "tractor" for a lorry employed to pull or tow anything. Thus all artillery prime movers are designated as tractors.

Loss replacement.—A replacement to fill a vacancy which has been created by the loss to the organization of the original occupant. A loss replacement differs from a *filler replacement*, which is a person added to a newly organized unit to bring it to its prescribed strength. *British equivalent*: **Reinforcement.**

Low-angle fire.—Gunfire delivered at angles of elevation below the elevation that corresponds to the maximum range of the piece, so that the ranges increase with increases in angles of elevation.

Low explosive.—A relatively slow-burning explosive, usually set off by hear or friction. Low explosives are therefore suitable for propelling charges in guns of for ordinary blasting, while **high explosives** are suitable of use in bombs and projectiles.

Low order detonation.—An incomplete and comparatively low explosion of the burster of a projectile.

Low wire entanglement.—A low barrier consisting of flat networks of barbed wire sloping down both ways from a line of stakes about three feet high; low entanglement.

Lung irritant.—A chemical agent which causes irritation and inflammation of the bronchial tubes and lungs.

Machete.—A large broad-bladed knife, sometimes two or three feet long, used chiefly for cutting paths through jungle.

Machine gun.—A gun that fires small-arms ammunition automatically and is capable of sustained rapid fire. It is usually fired from a mount.

Machine-gun nest.—A machine-gun emplacement, generally in a concealed and fortified position.

Magazine.—*a.* A case or changer for carrying cartridges inserted in, or attached to, a repeating gun. The cartridges are fed into the gun from the magazine.

   b. A building or other structure in which ammunition or explosives are stored.

Main artery.—*British term*: A concentrated system of communications from rear to front on which are established signal officers and signal centers.

Main artery of supply.—*British term*: see **Main supply road**.

Main attack.—That part of the attack where the commander concentrates the greatest possible power. *Compare* **Holding attack**. *British equivalent*: Same.

Main base.—*British term*: When a force is partly maintained from outside the actual theater of operations, that part of the base outside the theater. See also: **Forward base**.

Main body.—The principal part of a command or a formation. It does not include detached elements of the command, such as advance guards, outposts, connecting files, etc.

Main effort.—In each tactical grouping, the employment of the mass of the available means in a decisive direction. *British equivalent*: Same.

Main guard.—*British term*: see **Reserve**.

Main line of resistance.—A line joining the forward edges of the most advanced organized defense area. It is a series of separate centers of resistance which support each other. On this line the first determined effort is made to stop the enemy. *British equivalent*: *(Line of) forward (foremost) defended localities*.

Main supply road.—The principal inbound road over which supplies are carried to troops in the forward area. (Formerly called *axial road*.) *British equivalent*: *Main artery of supply*.

Maintenance, first echelon.—Driver's maintenance, covering the simple operations that can be trusted to the skill of the average driver using tools and supplies available on the vehicle. *British equivalent*: *First-line repairs* (i.e., daily maintenance by driver).

Maintenance, fourth echelon.—That maintenance normally performed in the rear areas by quartermaster and ordnance personnel. *British equivalent*: *Base repairs* (i.e., maintenance carried out by base workshops).

Maintenance requirements.—Those supplies required to replace expenditures. *British equivalent*: *Supplies* (classified as R.A.S.C., ordnance, RE, etc.).

Maintenance, second echelon.—That maintenance, other than first echelon maintenance, performed by the using arms and services. *British equivalent*: S*econd-line repairs* (i.e., maintenance carried out by light aid detachments and divisional workshops).

Maintenance, third echelon.—That maintenance normally performed in the field by quartermaster and ordnance personnel. *British equivalent*: *Third-line repairs* (i.e., maintenance carried out by army workshops).

Major.—An officer in the Army who ranks next above a captain and next below a lieutenant colonel. A major usually commands a battalion or an Army Air Forces squadron.

Major general.—An officer in the Army who ranks next above a brigadier general and next below a lieutenant general. A major general usually commands a division or an Army Air Forces command.

Major repair.—Repair work on items of matériel or equipment that need complete overhaul or substantial replacement of parts, or that require special tools.

Malingerer.—A person who pretends to be sick so that he can escape work or duty.

Maneuver.—Movement so designed as to place troops, material, or fire in favorable strategic or tactical locations with respect to the enemy; also a tactical exercise executed on the ground or map, in simulation of war and involving two opposing sides, though one side may be outlined, represented, or imaginary. The plural of the term applies to a series of such exercises, generally involving large bodies of troops in the field in simulation of war. *British equivalent*: **Manoeuver.**

Maneuvering force.—An element of a combat unit that carries out the main attack while the enemy is held by a secondary attack force; called pivot of maneuver; maneuvering element; mass of maneuver.

Maneuvers.—War games; a series of tactical exercises usually carried out by large bodes of troops in imitation of war.

Maneuver unit.—A command or subdivision of a command that is maneuvered as a unit in a tactical operation.

Map.—A representation (usually on a flat surface) of the surface of the earth, or some part of it, showing the relative size and position, according to some given scale or projection, of the parts represented. *British equivalent*: Same.

Map exercise.—An exercise similar to a *map maneuver* except that it is always one-sided, with students directing only the movement of friendly troops.

Map maneuver.—An exercise in which a military problem, or a series of problems, is solved on a map. In a map maneuver, the students may be divided into two groups, each directing the movements of a force, represented by markers, or they may direct only a friendly force, the instructor making the necessary enemy moves to develop the situations.

Maps, intermediate scale (large scale) (medium scale) (small scale).—See *Intermediate scale (large scale) (medium scale) (small scale) maps.*

Map scale.—The relation between the distance shown on a map and the actual distance on the ground. For example, one inch on a map drawn to a scale of 1:20,000 represents 20,000 inches, or 555.55 yards on the ground.

March.—*a.* To move in a steady, regular manner and in a given order, on foot, mounted, or in a vehicle.

b. A movement of troops from one place to another in this way.

c. A command of execution for troops to march in a given direction.

March collecting point.—A location on the route of march at which casualties who cannot continue to march are given medical treatment and are moved to medical stations in the rear.

March column.—One or more march units, or serials, marching on the same route under the control of a column commander.

March discipline.—The observance and enforcement of the rules of good marching, especially as relates to the conduct of individuals and operation vehicles. *British equivalent:* Same or **track discipline**.

March graph.—A graphical presentation of a march, used in planning and controlling marches and in preparing and checking march tables. *British equivalent:* **Movement graph**.

March on.—March toward. *British equivalent:* Same.

March outpost.—A temporary outpost established for the protection of the command during a brief halt, or while regular outposts are being established. *British equivalent:* **Covering detachment**.

March table.—A combined location and movement schedule for a march. *British equivalent:* Same or **movement table**.

March unit.—A subdivision of a marching column which moves and halts at the command or signal of its commander. *British equivalent:* **Group**.

Mark.—Letters or symbols formerly stamped on an item of matériel or equipment to show its design and model. In 1925, a system of **model numbers** was introduced to replace the mark for U.S. matériel and equipment. The British continued to use the mark for its equipment during World War II.

Marks (or Markings).—*British term:* see **Identifications**.

Marshalling yard.—*British term:* A place where loaded railway trucks (cars) arriving from various depots are formed into pack trains.

Martial law.—Military authority substituted for civil government in the home country or any district thereof, either by proclamation or as a military necessity, when the civil government is temporarily unable to exercise control. *British equivalent:* Same.

Marking panel.—A sheet of cloth or paper displayed by ground troops to signal their position and progress to a friendly aircraft, which then reports to headquarters. Marking panels are used only in the front lines.

Martial law.—Military authority substituted for civil government in the home country or any district thereof, either by proclamation or as a military necessity, when the civil government is temporarily unable to exercise control.

Mask (obstruction).—Any natural or artificial obstruction which interferes with the view of fire; usually an intervening hill, woods, etc. Friendly troops located between a gun and its target may constitute a mask. *British equivalent*: None.

Mass.—*a.* A concentration of combat power.

    b. To concentrate or bring together in one place; as, to mass the fires of all batteries.

    c. A military formation in which units are spaced at less than the normal distances and intervals.

Massed formation.—An arrangement of troops or units in a compact group with little space between them.

Mass of maneuver.—See ***Maneuvering force***.

Master sergeant.—A noncommissioned officer of the first grade in the Army, who ranks above a technical sergeant. A master sergeant is equivalent in rank to a first sergeant.

Matériel (or materiel).—Supplies, stores, and equipment of all types used in combat, including instruments, vehicles, clothing, weapons, and ammunition. Matériel and personnel are the two subdivisions of military power. *British equivalent*: ***Material***.

Matériel evacuation.—The transportation by service units of recovered matériel which has been severely damaged, from recovery collection points, on the axis of evacuation, or maintenance establishments, to ensure the eventual return of this matériel for further service or its use as scrap.

Maximum level of supply.—The maximum quantity of supplies authorized to be on hand and due in at any one time at a given supply point, defense command, department, theater of operations, or similar activity, measured in days of supply or in specific quantities of an item. The maximum level of supply is the ***minimum level of supply*** plus the ***operating level of supply***. For ammunition, it will include the total quantity authorized to be on hand in a particular command, including quantities in the hands of troops.

Means** of **signal communication.—An agency of signal communication capable of transmitting messages such as messenger, pigeon, radio, visual, sound, and wire communication. *British equivalent*: Same.

Mechanization.—A term used to denote the process of equipping a military force with armed and armored motor-propelled vehicles, such as tanks and combat cars. Mechanization differs from motorization, in that the ***motorization*** of a unit provides a means of transportation only, whereas in mechanization the unit both travels in, and fights from, its vehicles.

Mechanized cavalry.—Cavalry, equipped with, and fighting from, armed and armored motor vehicles instead of horses. *British equivalent*: Same.

Mechanized elements.—Those elements of cavalry equipped with armored and self-propelled motor vehicles designed for combat purposes and in which weapons are mounted.

Mechanized unit.—A unit which moves and fights in motor vehicles, the bulk of which are armed, and armored vehicles self-contained as to crew and weapons. *British equivalent*: Same.

Medium artillery.—*a.* Artillery pieces of medium size or caliber, usually ranging from 106-mm to 155-mm and including the 155-mm howitzer. The other two classes are ***light artillery*** and ***heavy artillery***.

b. Artillery units that use such pieces.

Medium machine gun.—*British term*: see ***Heavy machine gun***.

Medium scale maps.—Maps normally of a scale from 1:50,000 to 1:125,000, intended for strategical, tactical, and administrative studies by units ranging in size from the corps to the regiment.

Meeting engagement.—A collision between two opposing forces, each of which is more or less unprepared for battle. *British equivalent*: ***Encounter or contact battle***.

Meeting point.—*a.* The point at which guides meet their units or transportation.

b. British equivalent: A place at which third and second line transport is met by guides and directed to delivery points; a rendezvous where R.A.S.C. transport is met by unit guides and guided to unit transport areas where loads are transferred; see also ***Delivery point***.

Mess.—*a.* A section with any Army organization that prepares and serves food.

b. A group of officers or men who take their meals together.

Message.—A term which includes all instructions, reports, orders, documents, photographs, maps, etc., in plain language or code, transmitted by a means of signal communication. *British equivalent:* Same or ***signal*** or ***despatch.***

Message center.—The agency of the commander at each headquarters or command post charged with the receipt, transmission, and delivery of all messages except those transmitted directly by the writer to the addressee by telephone or personal agency, those handled by the military or civil postal service, local messages, and those arriving by special messengers. *British equivalent:* ***Signal centre*** or ***Signal office.***

Message, deferred (priority) (routine) (urgent).—See ***Deferred (priority) (routine) (urgent) message.***

Message in cipher (or in code).—*British term:* see ***Secret text (or secret language).***

Meteorological message.—A message in code giving data about atmospheric conditions for the use of artillery units; metro message.

Meteor report.—*British term:* A report and forecast of weather conditions issued to formations and units to assist them in planning operations, e.g. artillery shoots, smoke screens, gas attacks.

Method of capabilities.—A method of planning an operation based on a study of action open to the enemy. In using the method of capabilities, all lines of action open to the enemy are examined until a decision is made as to which line of action the enemy will probably use, or as to the order of probability of the various lines of action open to the enemy.

Method of intentions.—A method of planning an operation based on an estimate of what the enemy intends to do. In using the method of intentions, a knowledge of the arrangement, number, and kind of enemy forces is used to estimate what the enemy intends to do. When such knowledge is lacking, the method of intentions is used to estimate what action would be most advantageous for the enemy.

Metro message.—See ***Meteorological message.***

Microfilming.—Recording letters, drawings, or any documents in reduced form on photographic film. The film provides a permanent record, reduces bulk for transportation or filing, and permits inexpensive duplication. ***V-mail*** is usually microfilmed for shipment, and enlarged and printed for delivery.

Mil.—A unit by which angles can be measured. A mil is approximately the angle that is found when two lines making it are extended for 1000 yards and when the space separating them at that distance measures one yard. An ***artillery mil*** is an American unit for measuring angles, equal to 1/6400 of a circle. It is slightly smaller than the infantry mil; 100 artillery mils = 98.2 infantry mils. An ***infantry mil*** is a slightly larger unit, being the angle subtended by 1 yard at 1000 yards distance. 100 infantry mils = 101.8 artillery mils.

Military.—*a.* Of or related to the Army.
    b. Of the armed forces; related to the Army, Navy, etc.
    c. Having to do with war.
    d. The Army; soldiers.

Military attaché.—An Army officer on the official staff of an ambassador or minister to a foreign country. He serves as a military observer and reports to his own government on the military plans and developments of the country in which he is stationed.

Military channel.—A route of official communication between headquarters or commanders of military units. Usually called ***channel.***

Military commission.—A military court set up to try persons not normally subject to military law who are charged with military offenses. In places under military government or martial law, the military commission tries persons charged with either civil or military offenses.

Military correspondence.—Official letters, memoranda, reports, indorsements, telegrams, etc., of a military nature, usually sent between headquarters.

Military court.—A court of justice composed of military personnel that enforces military law; court; military tribunal. There are four kinds of military courts: military commission, provost court, court-martial, and court of inquiry.

Military courtesy.—Rules of conduct that are required, either by regulation or by tradition, for military personnel. Saluting officers is an important part of military courtesy.

Military crest.—The line nearest the crest of a ridge or hill from which all or nearly all of the ground toward the enemy and within range may be seen and reached by gunfire. The military crest is not always the ***topographical crest,*** which is the highest crest. *British equivalent: **Crest**.*

Military Forwarding Service.—*British term*: A service that works under **movement control** for the reception and forwarding of small consignments of equipment, clothing, instruments, gun sights, spare parts of motor vehicles, medical stores, private parcels, and comforts. It is also responsible for the safe disposal of the kits of sick, wounded, and deceased.

Military government.—A government established by the land or naval forces in enemy territory or in domestic territory recovered from rebels treated as belligerents. *British equivalent:* Same.

Military information.—Information, gathered from any source, which may serve to throw light on the enemy or the theater of operations. *British equivalent:* Same.

Military intelligence.—Evaluated and interpreted information concerning a possible or actual enemy, or theater of operations, together with the conclusions drawn therefrom. *British equivalent:* Same.

Military grid.—A network of straight north-south and east-west lines put on a military map, dividing it into squares to permit accurate location of points or places; grid. The distance between lines on the map represents a distance on the ground of 1000 to 100,000 yards, depending on the scale.

Military police.—A class of troops charged with the enforcement of all police regulations in the theater of operations and in other places occupied by troops. *British equivalent:* Same, commonly known as *"Red Caps"*.

Military rank.—The status held by military personnel that empowers them to exercise command or authority over other persons in the military service. Rank is divided into degrees or grades that mark the relative positions and powers of the persons holding it. Also called **rank.**

Militia.—An army of citizens trained for war or other emergency; local military defense organization. The National Guard is the organized militia of the United States; all other militia units are known as unorganized or reserve militia.

Mine.—*a.* A container holding an explosive charge that is put under water, laid on the ground, or buried. A mine is exploded at will by a control device or by contact with a vehicle, ship, etc. Mines are sed to hamper the movements of an advancing force or to make sea channels unsafe for enemy shipping. Sometimes a land mine is filled with a chemical agent rather than an explosive.

b. To put mines in a field of battle, waterway, etc.

c. To dig under enemy positions.

Mine field.—*a.* Area in which mines have been planted, such as a submarine mine field or a land mine field.

    b. A pattern of mines planted in such a field.

Mine mortar.—A small mortar or other device that throws projectiles containing high explosives for a short distance.

Minimum level of supply.—The minimum quantity of supplies to be held at a given point, base, defense command, department, or theater of operations, or similar activity, measured in days of supply or in specific quantities of an item. This amount should be held as a reserve and drawn against only in case of emergency.

Minimum range.—The least range setting at which the projectile will clear the mask when the gun is fired from a given position. *British equivalent*: **Minimum crest clearance.**

Misfire.—*a.* A failure to fire or explode properly.

    b. A failure of a primer or the propelling charge of a projectile to function, wholly or in part. Misfire can be contrasted with **hangfire**, which is a delay in any part of a firing charge.

Missing.—The classification of military personnel thought to be dead, although no positive evidence of death exists. The term is not limited to casualties in battle, but may include disappearances under unusual circumstances, such as in aircraft or vessels that have disappeared.

Missing in action.—*a.* The classification of military personnel thought to have been killed in combat, although the body of the soldier has not been found and no positive evidence of his death has been found.

    b. Thought to have been killed in combat, although no positive evidence of his death has been found.

Mission.—A specific task or duty assigned to an individual or unit or deduced from a knowledge of the plans of the immediate superior. For the Air Corps: each separate flight operation of a single airplane or of a formation. *British equivalent*: Same *or* **task.**

Mobile defense.—A defense of an area, position, etc., with combat weapons, in which maneuver is used together with the organization of fire and the utilization of terrain to seize the initiative from the enemy. Mobile defense if a form of **active defense.**

Mobile reserves.—*a.* Reserve troops who reinforce any defense that is threatened by the enemy.

    b. Reserve supplies held on trucks or on railroad cars for prompt movement forward. *British equivalent:* Same.

Mobile warfare.—Warfare of movement in which the opposing side seeks to seize the initiative by the use of maneuver, organization of fire, and utilization of terrain. Also called *war of movement.*

Mobilization.—The assembling and organization of troops, matériel, and equipment for active military service, in time of war or other national emergency. *British equivalent:* Same.

Mobilization plan.—A detailed plan for bringing the Army from its peacetime to its wartime establishment. This includes personnel, matériel, and equipment.

Mobilization point.—An area or camp at which troops are first assembled and given part of their training and equipment.

Model designation.—A mark, consisting of a combination of letters and numbers, stamped on an item of matériel or equipment to identify and classify it when it is adopted. The model designation shows what model the item is and how many changes, if any, have been made on the original model. M2A1 is an example, M2 showing the basic model, A1 showing a change in design.

Model number.—A number stamped on an item of matériel or equipment, when it is adopted, to show what model the item is, and how many changes, if any, have been made on the original model. M2A1, the 2 and 1 are the model numbers.

Mopping up.—The act of searching an area or position that has been passed over by friendly troops in the attack and of killing or capturing any enemy found. *British equivalent:* Same.

Morning report.—The daily report rendered to higher headquarters for the purpose of showing the status of individuals belonging to an organization. *British equivalent:* **Daily strength state.**

Morale.—The psychological condition or mental state of an individual or body of troops. *British equivalent:* Same.

Mortar.—An artillery weapon that has a relatively short barrel and generally a smooth bore. It has a shorter range and higher angle of fire than a howitzer, and is therefore used to reach targets that are protected by intervening hills or other short-range barriers.

Mortar carrier.—A cart of vehicle on which a mortar is mounted or transported. A mortar mounted in a half-track carrier is usually fired from the carrier.

Mosaic.—An assembly of two or more overlapping vertical aerial photographs: classified as "controlled," "uncontrolled," or "strip." *British equivalent*: Same.

Most immediate message.—*British term*: see **Urgent message.**

Motor convoy.—Transportation of military personnel or matériel by motor truck caravan, generally under escort.

Motorization.—The process of equipping a military force exclusively with motor-propelled vehicles. Motorization differs from **mechanization**, in that the motorization of a unit provides a means of transport only, whereas in mechanization the unit both travels in, and fights from, its vehicles, which are armed and armored. *British equivalent*: Same.

Motorized division.—An infantry division equipped with motor transport vehicles that enable its personnel, weapons, and equipment to be moved at the same time.

Motorized infantry.—Foot soldiers transported to and from battle fronts in motor trucks.

Motorized unit.—A unit equipped either organically or temporarily with sufficient motor vehicles to carry all its matériel and personnel at the same time. *British equivalent*: Same or **motor unit.**

Motor park.—*a.* See **Park**
   b. British term: see **Holding and reconsignment point.**

Motor pool.—Assembly of all vehicles of an organization into one group to be dispatched under the direction of the commander or his representative.

Motor repair park.—An area in the theater of operations to which motor vehicles are brought for repairs.

Motor transport.—Motor vehicles used for transport only, as contrasted with combat vehicles.

Motor transport pool.—A group of motor vehicles that are used in common by several units or that may be assigned to any particular organization whenever the need arises. Usually called **motor pool.**

Motor unit.—*British term*: see **Motorized unit.**

Motor vehicle.—A vehicle run with a motor; any wheeled, track-laying, or half-track vehicle that is powered by a motor. A trailer is also considered a motor vehicle.

Mountain artillery.—Light artillery that can be carried on pack horses or mules: artillery designed for use in mountainous country.

Mountain troops.—Soldiers equipped and trained in mountain warfare, including skiing and mountain climbing.

Mountain warfare.—*a.* Combat operations that are carried out in mountainous country.

    b. The tactics and techniques of fighting in mountainous country, as opposed to jungle warfare, trench warfare, etc. It includes skill in skiing, mountain climbing, etc.

Mounted.—*a.* Transported by vehicles or horses.

    b. In the saddle; on a horse; in a vehicle.

    c. Fastened onto a mount, backing, or support; especially, assembled on a gun mount.

Mounted defilade.—A position behind a natural or artificial obstacle high enough to give concealment and possible protection to mounted men.

Movable obstacles.—*British term:* see **Portable obstacles.**

Moving barrage.—*British term:* see **Rolling barrage.**

Movement.—Maneuver; the moving of troops and equipment from one place to another, as in a march or in transport overseas.

Movement by march route.—*British term:* see **Troop movement by marching.**

Movement control.—*British term:* A section of the Deputy-Quarter-Master-General staff responsible for the complete coordination of all movement. This is a joint organization of the Quarter-Master-General's branch and the Transportation Directorate. These work through Embarkation Staff Officers at ports, Railway Transport Officers at railheads, and Military Forwarding Officers. Administrative and technical working of the various transportation services is the responsibility of the Director General of Transportation.

Movement graph.—*British term:* see **March graph.**

Movement order.—An order issued to cover all the details of the movement of a unit, such as starting time, initial point, route to be followed, equipment to be taken, and refueling points for vehicles.

Movement to new position.—*British term:* see **Displacement.**

Move out.—A command that follows spoken field orders. It indicates that the men addressed are to leave and carry out orders.

Moving screens.—Patrols, often cavalry or mechanized detachments, used to keep enemy scouting parties at a distance and hide a body of moving troops from enemy observation.

Multiplace fighter.—A fighter airplane that has room for a crew of two or more men, in contrast with a **single-place fighter,** which is operated by a single flyer. It is used for patrol and escort duty, and also for attacking enemy ground positions and formations.

Multiple call.—*British term:* see **Conference call.**

Munitions.—Ammunition, explosives, and all other types of necessary war materials.

Munitions officer.—An officer in charge of the records and supply of ammunition for a unit.

Musette bag.—A small bag with a strap so that it can be carried over one shoulder, used by a soldier in the field to hold his own supplies and equipment; field bag.

Musketry.—*a.* Training in the use of the rifle and similar small arms.
   b. Fire delivered with rifles or similar small arms.

Mustard gas.—A very poisonous, oily, brown liquid that attacks the eyes and lungs and raises blisters on the skin; mustard. It has an odor similar to garlic, mustard, or horseradish. It is not actually a gas, but is spread in clouds made up of very small droplets.

Muster—*a.* To assemble for orders, review, inspection, etc.
   b. The assembling of troops for review, roll call, etc.

Mutiny.—*a.* A revolt or rebellion against military authority. It is different from sedition, which is a revolt against civil authority.
   b. To rebel or revolt against military authority.

Mutual support.—The support involving fire or movement or both, rendered one another by adjacent elements. *British equivalent:* Same.

Muzzle blast.—A rush of hot air and gases that bursts from the muzzle of a gun as the projectile leaves it. Sometimes called **powder blast.**

Muzzle flash.—A spurt of flames that appears at the muzzle when a gun is fired. It is caused by gases from the propelling charge that collect in the muzzle and ignite when mixed with air.

Muzzle velocity.—The speed of a projectile at the instant it leaves the muzzle of a gun; initial velocity.

National Guard.—A reserve part of the U.S. national military establishment that is administered by the States, and by some Territories. The National Guard is organized, armed, and equipped, wholly or in part, at federal expense, in peacetime. In time of war or national emergency it is called or ordered into federal service. In peacetime the National Guard acts as state guards.

National Guard of the United States.—An organization consisting of members and units of the National Guard who have taken an oath and been appointed for federal service whenever it comes necessary. The National Guard of the United States is at all times a component of the Army of the United States.

Natural obstacles.—Any terrain features which hamper military maneuvers or operations, such as deserts, mountains, streams, swamps, forest, etc. *British equivalent*: Same.

Navy Discipline Act.—*British term*: see **Articles of War**.

Net.—*a*. A radio system consisting of a number of stations in communication with one or another.

   b. A flexible barrier of steel mesh used to block entry to waterways, protect against torpedoes, etc.

   c. Fishnet, chicken wire, or similar foundation on which camouflage material is hung to conceal a gun, grounded aircraft, etc.

Neutralising fire.—*British term*: see **Neutralization fire**.

Neutralization.—*a*. Action taken to cancel, balance, or limit the effectiveness of an enemy area, weapon, installation, or force.

   b. Making a chemical agent harmless with a different chemical agent; as, the neutralization of mustard gas with bleaching powder. *British equivalent*: **Neutralisation**.

Neutralization fire.—Fire delivered for the purpose of causing severe losses, hampering, or interrupting movement or action, and in general destroying the combat efficiency of enemy personnel. *British equivalent*: **Neutralising fire**.

Neutralize.—To destroy or reduce the effectiveness of personnel or matériel by the application of chemicals or fire. *British equivalent*: **Neutralise**.

Nitrocellulose.—A chemical substance formed by the action of a mixture of nitric and sulfuric acid on cotton or some other form of cellulose. **Guncotton**, an explosive, is a nitrocellulose that has a very high nitrogen content.

Nitrocellulose powder.—A high-powered, smokeless propellent powder. **Pyro powder** is a type of nitrocellulose powder.

Nitroglycerin.—A powerful explosive oily liquid made by treating glycerin with a mixture of nitric and sulfuric acid.

No man's land.—A strip of land between the front lines of two opposing forces.

Noncombatant.—A person or organization whose duties do not involve actual fighting or the bearing of arms. Chaplains and members of the Medical Corps are noncombatants.

Noncommissioned officer.—An enlisted man holding any grade from corporal or technician, fifth grade, to, and including, master sergeant or first sergeant.

Nonpersistent agent.—A chemical agent whose effectiveness in the air at a point of release is dissipated within 10 minutes. *British equivalent:* Same (but with no definite time limit).

Nontoxic.—Not poisonous. *British equivalent:* Same.

Normal barrage.—A standing barrage laid in immediate defense of the sector which it supports. The barrage which is fired on prearranged signal from the support unit. *British equivalent:* **Barrage.**

Normal supply.—*British term: see* **Automatic supply.**

Normal zone.—That portion of the zone of fire of a unit within which its fire is ordinarily delivered. *British equivalent:* **Zone of fire.**

Nose spray.—Fragments of a bursting shell that are thrown out to the front in the line of flight in contrast with **nose spray,** which are fragments of a bursting shell that are thrown out to the rear, and **side spray,** which are fragments of a bursting shell that are thrown to the side.

Notice.—*British term:* The period of time, laid down by a superior commander, that will be available for preparation after units have been named for action and before they are required to move. The length of time governs the state of readiness. Thus, "one hour's notice" means that the unit of formation must be able to move one hour after it receives its order to do so.

Objective.—A locality which a command has been ordered to reach and occupy or a hostile force which a command has been ordered to overcome. For the Air Corps: That locality or thing which must be destroyed in order to accomplish an assigned mission. *British equivalent:* Same.

Oblique aerial photograph.—An aerial photograph made with a camera whose optical axis is oblique.

Observation aviation.—Units whose primary functions are reconnaissance and observation of near objectives, observation of artillery fire, and command, courier, and liaison duty for ground units. *British equivalent:* **Reconnaissance aircraft** (used to carry out both strategic and tactical reconnaissance).

Observation point.—*British term: see* **Cossack post.**

Observation post.—A point selected for the observation and conduct of fire, for the observation of an area or sector, for the study of objectives, or for the purpose of securing information of the enemy and his activities. A position from which friendly and enemy troops can be seen and from which fire is controlled and corrected. *British equivalent:* Same.

Observed fire.—Fire which can be adjusted by ground observation of the target either at the emplacement of a weapon or at an observation point in liaison with it.

Obstacle.—Any device or feature, either natural or artificial, used in field fortifications for the purpose of delaying the hostile advance. A natural terrain feature or artificial work which impedes the movements of the troops. Obstacles are classified as natural or artificial, tactical or protective, fixed or portable, etc. *British equivalent:* Same.

Offensive defense.—A defense consisting of attack, or of the active use of troops and fire power so as to protect against enemy attack; offensive-defensive.

Offensive-defensive.—See *Offensive defense.*

Officer of the day.—An officer having general charge of the interior guard and prisoners for a particular day. *British equivalent:* **Orderly officer.**

Open column.—A lengthened or expanded column of moving vehicles or units with long distances between them. An open column is a formation used to lessen the effect of enemy air attack, or to cut down crowding and collision within a column.

Open flank.—A flank of a unit in combat that is not protected by another unit or by terrain features, and that is exposed to attack.

Open formation.—An arrangement of men, aircraft, or mechanized units with space enough between them to allow some freedom of action or to lessen the effect of air or artillery attack.

Operating level of supply.—The quantity of supplies, measured in days of supply or in specific quantities of an item, necessary for the maintenance of a command. This level is determined frequency of shipments and the time required for supplies to reach their destination.

Operation instructions.—*British term:* see **Fragmentary orders** and **Letters(s) of instruction.**

Operation map.—A graphic presentation of all or parts of a field order, using conventional signs, military symbols, abbreviations, and writing or printing.

Operation order.—*British term*: see **Combat orders** and **Field order.**

Operations.—A complete process of carrying on combat on land or sea, including movement, supply, attack, defense, and all maneuvers needed to gain the objectives of any battle or campaign.

Oral order.—An order delivered by word of mouth. *British equivalent:* **Verbal order.**

Orderly.—*a.* A soldier who assists an officer, usually by carrying messages and other orders and by performing routine duties. *British equivalent*: Same. See also *British term* **Batman.**

b. An attendant in a mess, hospital, etc. *British equivalent*: Same.

Orderly officer.—*British term*: see **Officer of the day.**

Order of march.—The disposition of troops for a march, or their order in the march column.

Orders (or "O") group.—*British term*: see **Group.**

Ordnance.—Military matériel, such as combat weapons of all kinds, with ammunition and equipment for their use, combat and special purpose vehicles, and repair tools and machinery.

Ordnance officer.—*a.* Any officer who is a member of the Ordnance Department.

b. A special staff officer who advises his commander on technical questions of ordnance. An ordnance officer also has charge of the repair and maintenance of armament and the supplying of ammunition. In this meaning, also called ordnance staff officer.

Organic.—Assigned to, and forming an essential part of, a military organization. Organic parts of a unit are those listed in its **Table of Organization.**

Organizational maintenance.—The routine preventive care and adjustment of vehicles and equipment by the unit using the equipment.

Organizational requirements.—Those supplies necessary for the organization to function as a unit. *British equivalent*: **Requirements.**

Organizational unit loading.—See **Unit loading.**

Organization for combat.—The measures taken by a commander to insure that the troops of his command are so grouped that they can most efficiently carry out the mission assigned. *British equivalent*: None.

Organization of the ground.—The development of existing ground features to strengthen a defensive position, by the use of camouflage, field fortifications, etc.

Organized position.—An area in which troops and weapons have been put in position for future action, and in which field fortifications have been constructed.

Organized Reserves.—One of the peacetime components of the United States Army, made up of the Officers' Reserve Corps, the Enlisted Reserve Corps, and some officers and men of the Regular Army. It is organized and maintained in peacetime for purposes of military training, and as a source of trained military personnel in the event of a national emergency.

Orient.—To determine one's position on the ground with respect to a map or to the four cardinal points of the compass. To identify directions on the terrain. To place a map so that its meridian will be parallel to the imaginary meridian on the ground, and all points on the map in the same relative positions as the points on the ground which they represent. To inform or explain, to make another conversant with. *British equivalent*: Same.

Other ranks.—*British term*: In the British Army, includes warrant officers, non-commissioned officers, soldiers, apprentices, and boy entrants. See also ***Airmen***, ***Enlisted man***, and ***Ratings***.

Outflank.—To pass around or turn the flank or flanks of an enemy. To extend beyond the flanks of the enemy's line. *British equivalent*: Same.

Outguard.—The most forward security unit posted by an outpost. For example, a detachment of at least sufficient strength to post three reliefs of single or double sentinels. Usually posted by a support and consists of a squad or half squad. In larger outposts, a rifle platoon may be posted as an outguard. Where a support posts more than one outguard, outguards are numbered from right to left within each support. See ***Support (of outpost)***. *British equivalent*: ***Forward standing patrol***.

Outpost.—A detachment detailed to protect a resting or defending force against surprise by hostile ground forces. The outpost takes up its position at some distance from the main body of the command. See ***Combat outpost***. *British equivalent*: Same.

Outpost area.—A belt of terrain lying in front of a battle position, occupied by the observation or outpost elements. *British equivalent*: **Line of outposts.**

Outpost line of resistance.—A line designated to coordinate the fires of the elements of the outpost and its supporting artillery. For example, where the outpost is ordered to hold its position in case of hostile attack, the outpost line of resistance has the same function in respect to the outpost as the main line of resistance has for the battle position. *British equivalent*: None.

Over.—A shot which strikes or bursts beyond the target or over the target. Overs and shorts are observed in *sensing*.

Overlay.—A sheet of translucent paper or cloth, for laying over a map, on which various locations, as of artillery, targets, field works, enemy positions, etc., are shown. *British equivalent*: **Trace.**

Overprint.—New material printed or stamped upon a map or chart to show data of importance or special use, in addition to that originally printed.

Overseas cap.—See **Garrison cap.**

Pace.—A step of 30 inches; the length of the full step in quick time; rate of movement. *British equivalent:* Same.

Pack artillery.—Field artillery that is transported by pack animals. **Mountain artillery** is the most common type of pack artillery.

Pack howitzer.—A small artillery weapon transported, usually in sections, on pack animals. The pack howitzer was designed for use by mountain troops, but it is also used on mobile gun mounts in armored units.

Pack section.—*British term*: That part of the supply train containing the day's needs of a particular formation, consisting not only of supplies (including liquid fuels and lubricants) but also the miscellaneous articles required for the daily maintenance and comfort of the troops, such as engineer, motor transport, and ordnance stores, and gifts, comforts, and mails.

Pack train.—A column of pack animals with all necessary equipment and personnel.

Panel code.—A set of symbols consisting of cloth strips or panels displayed on the ground. It is used by ground troops to signal to aircraft for air-ground communication.

Panzer grenadier.—German armored infantry.

Parachute troops.—Troops moved by air transport and landed by means of parachutes. *British equivalent*: Same.

Parade.—*British term*: see **Assembly.**

Parent unit.—An organization to which a smaller unit, or individual, belongs. The smaller unit or individual may be on special duty or detached service with another organization and still belongs to the parent unit.

Park.—An area used for the purpose of servicing, maintaining, and parking vehicles. *British equivalent*: Same.

Parlementaire.—An agent of a field commander sent openly within enemy lines to communicate and negotiate directly with the enemy commander.

Parley.—An informal conference between enemies under truce, to discuss terms, conditions of surrender, etc.

Partisan warfare.—Operations carried on by small independent forces, generally on the flanks and rear of a superior enemy; guerrilla warfare. Partisan warfare is carried on to cause delay, disruption of communication and supply lines, and repeated small losses to wear down an enemy.

Party.—A detachment of individuals employed on any kind of duty or special service. For the artillery: Certain key officers and men who usually accompany the commander on the march and assist him in reconnaissance, in issuing his initial orders, in initiating the movement forward to position, and in the occupation and organization of the position. *British equivalent*: Same.

Passage of lines.—A relief of a front line unit in the attack in which the rear unit moves forward through the already established line: the unit passed through may remain in position or move to the rear. *British equivalent*: **Leapfrogging.**

Passive defense.—Defense in place, designed only to keep the enemy out of a position or area; passive protection. The primary means of passive defense are dispersion, concealment, camouflage, cover, and obstacles. Passive defense is not expected to furnish an opportunity for taking the initiative, as is *active defense.*

Password.—A secret word or distinctive sound which identifies a person or party desiring to pass a sentinel or enter a guarded area or building. A password is used as an answer to a challenge.

Patrol.—A moving group or detachment sent out from a larger body on an independent or limited mission of reconnaissance or security or both; the act of patrolling. *British equivalent*: Same.

   Standing Patrol.—*British term*: A party of from two men to a troop, or even more, posted a considerable distance in advance of other troops, to watch either the enemy, a route by which he might advance, or a locality in which he might concentrate unseen.

Pattern bombing.—The systematic covering of a target with aerial bombs, according to a plan. Pattern bombing differs from **area bombing**, which is bombing of a general area according to no special plan, and from **precision bombing**, which is bombing at a specific target.

Paulin.—A long sheet of olive-drab canvas, usually treated to make it resistant to moisture and chemicals, used as a protective cover. A paulin is used to cover guns, etc., in the open, or to put underneath articles unloaded on damp ground.

Penetration.—A form of attack in which the main attack seeks to break the continuity of the enemy's front and to envelop the flanks thus created. *British equivalent:* Same.

Pennant.—A small triangular flat usually flown for identification of a unit or a general officer. A pennant with two stars on a red background indicates the presence of a major general.

Persistency.—The length of time a chemical agent will remain in an effective concentration after release in open air.

Persistent agent.—A chemical agent which will maintain an effective vapor concentration in the air at point of release for more than 10 minutes.

Personal clothing and equipment.—*British term:* see **Individual equipment.**

Personnel.—The body of people in a military force. Personnel and matériel are the two subdivisions of military power.

Personnel carrier.—A motor vehicle, sometimes armored, designed primarily for the transportation of personnel and their weapons to and on the battlefield. *British equivalent:* Same.

Personnel mine.—See Antipersonnel mine.

Petrol.—*British term:* gasoline.

Petrol point.—*British term:* A point where a forward holding on wheels is established by a second line R.A.S.C. company to facilitate the replacement of petrol (gasoline) and lubricants to forward units. See **Distributing point.**

Phase line.—A line or terrain feature on which units may be halted for control, coordination, or further orders. *British equivalent:* see **Report line.**

Phase of the attack.—A distinct stage of an attack. The phases of an attack usually include the approach march, deployment, fire fight, assault, reorganization, and pursuit.

Phonetic alphabet.—A list of stander words used to identify letters in a message give by radio or telephone, such as *baker* for *B* or *king* for *K*.

Platoon defense area.—*See **Defense area***. For example, a platoon defense area comprises a section of the company defense area assigned to a rifle platoon as its task in the all-around defense of the company area.

Phosphorus.—A chemical agent used to start fires. Phosphorus comes in two forms: red, which is nonpoisonous but highly inflammable and explosive when mixed with oxidizing chemicals; and white, which is extremely poisonous and highly inflammable. White phosphorus is often used to produce smoke.

Photographic reconnaissance.—All military aerial photography accomplished for other than mapping purposes. *British equivalent*: Same.

Photomap.—An aerial photograph upon which information commonly found on maps has been placed, including at least a scale and a directional arrow. *British equivalent*: None.

Picket.—A detachment of an outpost sent out to perform the duties of an outguard at a critical point, the detachment being stronger than an ordinary outguard and establishing sentinel posts of its own. A stake of wood or steel used in the construction of revetments and obstacles. To guard, as a camp or road, by an outlying picket. To post as a picket. *British equivalent*: **Piquet** or **standing patrol**.

Picket line.—A rope or cable stretched about four feet above the ground to which tie ropes on horses or mules may be fastened.

Piecemeal attack.—An offensive action in which the various units are employed as they become available, and not according to a plan for the effective use of the force as a whole, as in a **coordinated attack**.

Pillbox.—A small, low, fortification that houses machine guns, antitank weapons, etc. A pillbox is usually made of concrete, steel, or filled sandbags and is used as a point of resistance in defense.

Pin point.—A vertical aerial photograph in which the object of interest is centered.

Pioneer.—A person selected, trained, and equipped for rough engineering work in advance of the main body of troops.

Piquet.—*British term*: see *Picket.*

Plain text (or clear text) (or plain language).—The text of a message which, on its face, conveys an intelligible meaning in a spoken language.

Plan.—A scheme or design, specifically for any military operation. A course of action or method of procedure decided upon and adopted by a commander, and which is the basis for his orders to his command.

Plan of attack.—*see* **Scheme of maneuver.**

Platoon.—*a.* The basic tactical unit of the Army; a subdivision of a company, battery, or troop. A platoon is composed of two or more squads or sections.

   b. British: A quarter of an infantry company. Consists of four sections.

Plunging fire.—Gunfire that strikes the ground at a high angle. Plunging fire is less effective than grazing fire, which covers a longer zone and my cause damage outside of the zone actually hit.

Point.—The patrol or reconnaissance element which precedes the advance party of an advance guard, or follows the rear party of a rear guard. *British equivalent*: **Point section of the van guard.**

Point-blank range.—The distance to a target that is so short that the trajectory of a bullet or projectile is practically a straight rather than a curved line. Point-blank range is one for which no elevation is needed.

Point of maneuver.British term: see **Battle position.**

Point section of the van guard.—*British term*: see **Point.**

Poisonous gas.—*British term*: see **Casualty agent (chemical).**

Ponton (or pontoon).—A light boat or float used as one of the floating supports for a temporary military bridge, or as a raft to ferry items over water.

Portable obstacles.—Obstacles capable of being moved. *British equivalent*: **Movable obstacles.**

Portée (or porté).—*a.* Carried in, or towed by, vehicles. The term is used in connection with artillery pieces or cavalry units.

   b. To transport artillery pieces or cavalry units in vehicles.

Port of embarkation.—An army organization established for the purpose of administering and controlling the embarkation or dis-embarkation of troops and supplies at a supply point.

Position.—*a*. An area or locality occupied by combat elements, espe-cially for defense.

b. The location of a gun, unit, or individual from which fire is de-liver upon a given target. For supporting weapons in combat, he firing positions are known as primary firing position, alternate fir-ing position, and supplementary firing position. In this meaning, also called *fire position* and *firing position*.

Position defense.—A type of defense in which one main position of resistance is established in the defensive area. The main defense of the whole area is conducted from this position rather than from a series of locations, as in *zone defense*.

Position defilade.—A position of a gun from which the gun crew can see the target, but which is hidden from enemy observation by an obstacle, such as a crest of a hill; site defilade.

Position in observation.—*British term*: (Artillery) Implies batteries in action watching all ground in their field of fire and ready to open fire.

Position in readiness.—*a*. A position assumed as a temporary expe-dient in a situation so clouded with uncertainty that positive action is considered unwarranted. *British equivalent*: **Position of readi-ness.**

b. British: (Artillery) Implies batteries limbered up under cover with all possible alternative positions in the immediate neighbor-hood reconnoitered and everything ready for their occupation.

Position of resistance.—A position chosen by a commander as the point at which to stop and defeat the enemy; especially, a prepared position to which a retreating force withdraws.

Position warfare.—Warfare in which the defensive is confined main-ly to fixed positions, as contrasted with **war of movement**, which involves the use of maneuver. The defense is aimed chiefly at keep-ing the enemy out of strategic areas and forcing him to exhaust his combat power in assaults against well-established positions. Also called **war of position**.

Post.—*a*. A military installation or location at which troops are sta-tioned. A post may be a camp, depot, fort, hospital, proving ground, station, arsenal, air base, air field, etc. In this meaning, also called *garrison*.

b. An area for which a guard or sentry is responsible; location of a soldier while on active duty.

Post exchange (PX).—Exchange.—A military organization that sells merchandise and services to military personnel and other authorized personnel. Often called an *Army exchange* or *exchange*. British equivalent: *Canteen* or *Navy, Army, and Air Force Institutes (NAAFI)* (known abroad as *Expeditionary Force Institutes (EFI)*).

Powder.—An explosive compound or mixture that is used as a propelling charge or bursting charge in projectiles, bombs, mines, etc. There are two general types, *black powder* and *smokeless powder*.

Powder bag.—A fabric container that holds the powder charge for separate-loading ammunition. A powder bag is usually made of special silk or cotton cloth that burns without leaving any ashes.

Prearranged fires (or scheduled fires).—Supporting fires for which the fire data are prepared in advance and which are delivered to a time schedule or on call from the supported troops. *British equivalent: Predicted fire.*

Precede.—To regulate movement on the element in rear. *British equivalent*: Same.

Precision bombing.—Bombing with special sights and other aids so that the bombs are directed at a specific target with the greatest possible accuracy. Precision bombing differs from *area bombing*, the bombing of a general area according to no special plan, and from *pattern bombing*, the systematic covering of a target according to a plan.

Predicted fire.—Firing at the point at which a moving target is expected to be when the projectile reaches it, according to predictions based on observations (*Also British term*: see *Prearranged fire* (*or schedule fire*).)

Preliminary bombardment.—Concentrated and intensive artillery fire in preparation for an attack.

Preparation, artillery.—See Artillery preparation.

Primacord.—A flexible fabric tube containing a filler of high explosive PETN (pentaerythritetetranitrate) that is used as a bursting charge or as a primer for other high explosive charges. *Primacord* was the trade name for the type of *detonating cord* then in use.

Primary armament (Coast Artillery Corps).—Seacoast artillery weapons of 12-inch or greater caliber. *British equivalent: Superheavy coast defence guns* (approximately).

Primary firing position.—The position from which a unit or weapon executes its primary mission. Also called **primary fire position**. *British equivalent*: Same.

Primary target area.—The target area assigned as the principal fire mission of a weapon or unit. Compare **Secondary target area**.

Prime mover.—A motor vehicle used to tow a gun or trailer. *British equivalent*: **Tractor**.

Primer.—*a*. A device used to set off a propelling charge or a bursting charge. A primer may consist of an igniter and a sensitive charge. The igniter is set in action by friction, pressure, or electricity, and it in turn sets off the sensitive charge, which explodes the propelling charge or the bursting charge, or ignites the burning mixture of certain incendiary bombs.

    b. A small charge of high explosive fitted with a detonating cap or detonating cord. Priming a large charge consists of placing the primer in the main charge.

Principle supply road.—A supply road needed to supply an element or unit that is so located as to need a separate road; main supply road.

Priorities.—Definite rulings which establish, in order of time, the precedence of shipment; movements of rail, road, water, or other transport; or performance of several tasks.

Priority call.—*British term*: see **Urgent call**.

Priority message.—A message of less urgency than those entitled to urgent classification but which warrants precedence over routine messages in order to reach the addressee in time for effective action. *British equivalent*: **Important message**.

Prisoners of war.—Persons captured and held in captivity or interned by a belligerent power. *British equivalent*: Same.

Private.—A soldier in the lowest grade in the Army. *British equivalent*: Trooper (cavalry), Gunner (artillery), Sapper (engineers), Signalman (Signals), Driver (transport), Guardsman (Brigade of Guards), Fusilier (fusilier), Rifleman (rifle battalion), Aircraftsman (Royal Air Force), or private (other).

Procedure.—*British term*: A routine laid down for commanders, inter-communications and liaison personnel and vehicles to ensure that intercommunication, reconnaissance, and liaison are continually maintained.

Prolonge.—(pro LAWNJ). A rope, with a hook or loop at one end, with which soldiers can move a vehicle or gun carriage into position.

Prone position.—The posture of the body for firing from the ground.

Protection.—*British term:* see **Security.**

Protective clothing.—Clothing that is specially treated to protect the body of the wearer from coming into contact with chemical agents such as mustard gas.

Protective fires.—Fires placed by supporting weapons on enemy rearward areas for the purpose of hindering enemy fire or movement against the friendly attacking rifle or tank elements. For infantry weapons such fires are usually delivered at long range.

Protective obstacles.—Obstacles whose chief purpose is to prevent a sudden incursion of attacking forces. *British equivalent:* **Defensive obstacles.**

Proving ground.—A testing ground; place or area at which articles of matériel and equipment, especially artillery and motor vehicles, are tested to make sure they are in perfect working condition.

Provost court.—A military court which tries soldier or civilians who commit minor offenses within the limits controlled by the Army.

Provost guard.—Men on special duty, under the provost marshal, to keep order among soldiers when they are outside the territory under the control of the interior guard. The provost guard usually serves when military police are not available, and usually works in cooperation with nearby civil authorities.

Provost marshal.—A staff officer who commands the military police detachment of a post or organization. A provost marshal advises the commander on military police matters.

Pursuit.—An offensive operation against a defeated enemy for the purpose of accomplishing his annihilation. *British equivalent:* Same.

Pursuit aviation.—That type of aviation whose primary function is air fighting; classified as interceptor and fighter. *British equivalent:* **Fighter aircraft.**

Pyrotechnic pistol.—A pistol from which fireworks signals, especially flares, are fired. A pyrotechnic pistol is used in aircraft. Also called s**ignal pistol.**

Pyrotechnics.—Ammunition containing chemicals that produce brilliant light in burning, used for signaling or lighting up areas at night; fireworks.

Quartering.—Providing shelter for troops, headquarters, establishments, and supplies.

Quartering party.—A detail sent out to reconnoiter for billets or quarters. A billeting party. *British equivalent:* **Harbouring party**.

Quartermaster.—An officer of the Quartermaster Corps who serves as a staff officer of a unit or post and has charge of quartering, supply, etc.

Quartermaster agent.—A civil service employee who serves on a transport when a commissioned officer of the Quartermaster Corps is not available for duty. A quartermaster agent travels on the transport and attends to all the duties connected with the quarters and supplies for the troops.

Quartermaster Corps.—A branch of the Army Service Forces that is responsible for the provision of food, clothing, equipment, housing, etc. for the Army.

Quartermaster depot.—An establishment for receiving, storing, and issuing quartermaster supplies.

Quartermaster General, The.—An officer, with the rank of major general, who is the head of the Quartermaster Corps and is responsible for the direction and supervision of its services.

Quarters.—A place or structure in which troops are housed; lodging for soldiers.

Quickmatch.—A fast-burning fuze made from a cord impregnated with black powder, used in flares.

Quick time.—A rate of march at 120 steps, each 30 inches in length, a minute. It is the normal cadence for drills and ceremonies.

Radio net.—A system of military radio stations operating with each other. A radio net usually includes the radio station of the superior unit and those of all subordinate or supporting units.

Radiotelegraphy (or radio (key)).—Radio communication by means of the International Morse Code. *British equivalent:* **Wireless-telegraphy**.

Radiotelephony (or radio (voice)).—Radio communication by means of the voice. *British equivalent:* **Radio-telephony**.

Raid.—A sudden and rapid incursion. An offensive movement, usually by small forces directed against an enemy. *British equivalent:* Same.

Raiding party.—Troops that make a sudden quick attack on the enemy. Raiding parties may be sent out to take prisoners, get information, or harass the enemy.

Railhead (truckhead) (navigation head).—A supply point where loads are transferred from the particular type of transportation being employed; such as "Class I Railhead, 1st Division," "Ammunition Railhead, 1st and 2d Divisions." *British equivalent*: Same.

Railhead distribution.—Issue of Class I supplies to regimental (or similar unit) transportation at the railhead. *British equivalent*: None.

Railhead maintenance area.—*British term*: An area in the vicinity of railheads in which are situated advance depots and dumps for the maintenance of forward formations, served by these railheads, and of local units.

Railway artillery.—Artillery mounted on and fired from railway cars; railroad artillery.

Railway siding.—*British term*: see **Holding and reconsignment point**.

Rally.—*a*. To bring troops back together after a disorganized action; as, to rally retreating troops.

b. To come together after disorganizing action.

c. A position in the air to which a unit of aircraft returns after an action; rallying point of an aviation unit.

d. British:

(i) **Forward rally.**—An area under cover in the vicinity of a captured objective, to which tanks go when released from dominating the objective and in which they are immediately available to repel a counterattack.

(ii) **Rear rally.**—The area to which tanks go when released by the local commander from the forward rally.

Rallying point.—A point, designated by a unit commander, where he assembles his unit for further operations after the attack of an objective. For example, used by Infantry in night attacks and similar operations. *British equivalent:* Same.

Ranger.—A soldier specially trained to make surprise attacks on enemy territory. Rangers act in small groups, making rapid attacks and withdrawing. **Ranger** is used by the Americans; the corresponding British term for soldiers of this kind is **Commando**.

Ranging.—*a*. Correcting the range settings of a gun or battery by observation of fire.

b. Wide-scale scouting, especially by aircraft, designed to search an area systematically.

c. Locating an enemy gun by watching its flash, listening to its report, or other similar means.

Rank.—*a.* The status held by military personnel that empowers them to exercise command or authority over other persons in the military service. Rank is divided into degrees or grades that mark the relative positions and powers of the persons holding it. In the meaning, also called **military rank.**

b. To possess a higher rank than another.

c. A line of persons or thinks arranged side by side. Both **rank** and **file** refer to single lines of troops, vehicles, etc., in formation; **ranks** are lateral lines from side to side; **files** are vertical lines from front to rear.

Ranks.—Ordinary soldiers; enlisted men.

Rate of fire.—The number of shots per minute.

Rate of march.—The average speed over a period of time including short periodic halts. *British equivalent*: Same *or* **speed.**

Rating.—A specialist grade held by an enlisted man; specialist grades in the Army Air Forces held by officers or enlisted men, such as pilot, parachutist, navigator, gunner, etc.

Ratings.—*British term*: In the Royal Navy, enlisted men graded upwards from boy, 2nd class, to chief petty officer. See also **Airmen, Enlisted man**, and **Other ranks.**

Ration.—The prescribed allowance of the different articles of food for the subsistence of one person or one animal for one day. See also **Field ration** and **Garrison ration.** *British equivalent*: Same.

Ration cycle.—The period of time within which the three meals of a ration are consumed. Normally it begins with the evening meal. *British equivalent*: None.

Ration strength.—The estimated number of men entitled to rations on any day.

REACK.—Receipt acknowledged. REACK is a code abbreviation used in transmitting telegrams and teletype messages.

Rear.—That part of a force which comes last or is stationed behind the rest. The direction away from the enemy. *British equivalent*: Same.

Rear echelon.—That part of a headquarters engaged in administrative and supply duties and located a considerable distance behind the front lines. The **forward echelon** includes the staff agencies and command facilities that a commander needs in tactical operations. *British equivalent*: **Divisional administrative area** and **Divisional administrative group.**

Rear maintenance area.—*British term*: An area in which are located temporary depots which maintain the force ashore, from the time when issues from beach maintenance areas cease, and until permanent base depots start, to issue.

Rear guard.—A security detachment which follows the main body and protects it on the march.

Rear party.—The detachment from the support of a rear guard which follows and protects it on the march. *British equivalent*: Same or **rear patrol.**

Re-clothing point.—*British term*: A forward detachment of the petrol (gasoline) company, consisting of one or more lorries (trucks) from which units may draw reserve clothing to replace any that has become contaminated.

Recognition.—The process of determining by visual methods whether an object is friendly or not. The process of determining whether an object is friendly or not by other than visual means is called **identification.**

Reconnaissance.—The operation of searching for information in the field. Reconnaissance is characterized by the type of information sought, as in engineer reconnaissance; and by the method of seeking it, as in photographic or air reconnaissance. *British equivalent*: Same.

    a. The U.S. Army divided reconnaissance into five categories:

     Battle reconnaissance.—Continued observation made, under combat conditions, of the terrain, disposition of the enemy, etc. It is made during or immediately before battle, when in close contact with the enemy.

     Close reconnaissance.—Reconnaissance of a region close at hand. Close reconnaissance furnishes the commander with information upon which he makes his tactical decisions.

     Combat reconnaissance.—Reconnaissance of the enemy in immediate contact with one's own forces, preliminary to, or during, combat.

     Distant reconnaissance.—Exploration of objectives that lie outside immediate striking range of a force, but about which detailed information is essential for military planning.

     Strategic reconnaissance.—Search over wide areas, usually by air, to gain information of enemy concentrations or movements that would aid in making strategic or large-scale decisions.

b. The British Army divided reconnaissance into three categories:

Close reconnaissance.—Carried out at close quarters to the enemy to obtain detailed information of the ground, and of the enemy's location, strength, and movements.

Medium reconnaissance.—Carried out to obtain general information about the enemy's movements and disposition on which to base a plan of offensive or defensive operations.

Distant reconnaissance.—Carried out to discover the enemy's strategical movements and concentrations in a theater of operations.

Reconnaissance aircraft.—*British term*: encompasses both *Observation aviation* and *Reconnaissance aviation*.

Reconnaissance aviation.—Units whose primary function is reconnaissance of distant objectives.

Reconnaissance by fire.—The search for an enemy position by firing on his probable position and thus drawing fire.

Reconnaissance car.—An automobile used principally for scouting.

Reconnaissance echelon.—A unit or element engaged in reconnaissance. A reconnaissance echelon may be part of an armored regiment.

Reconnaissance element.—Any unit engaged in reconnaissance. It may be aviation, cavalry, tanks, motorcyclists, infantry in trucks, etc.

Reconnaissance (or "R") group.—*British term*: see *Group*.

Reconnaissance in force.—An attack by a considerable force of troops used to discover and test the enemy's position and strength.

Reconnaissance of position.—A detailed examination of terrain as a basis for the selection of advantageous locations for guns and troops.

Reconnaissance patrol.—A patrol whose primary mission is to obtain information, maintain contact with the enemy, or to observe terrain. *British equivalent*: Same or *mission (or area search) sortie*.

Reconnaissance strip.—A series of overlapping vertical photographs made from an airplane flying a selected course. *British equivalent*: *Line overlap* (vertical and oblique) or *mosaics* (vertical only).

Reconstitute.—To reestablish a unit either on the active or inactive list of the Army. This may be done only by the authority of the Secretary of War.

"Red Caps".—British term: see **Military police.**

Reduced charge.—Ammunition having a propelling charge less than normal.

Refilling.—*British term*: The process of unloading, breaking bulk, and reloading on a unit basis.

Refilling point.—*a.* A former name for **Supply point.**

b. British: (i) In the case of material other than ammunition, the place where articles hitherto carried in bulk are reloaded in detail for units In the case for ammunition, the place where loads are transferred from third line to second line transport.

(ii) A place where the transfer of loads from one detachment of R.A.S.C. lorries (trucks) to another takes place. It will only be used when the distance between the forward maintenance area and the delivery point is beyond the range of second line transport, or when forward formations are being maintained direct from the railhead maintenance area. It is in no sense a dump.

Regiment.—An administrative and tactical unit of the Army. A regiment is larger than a battalion, smaller than a brigade or division, and is usually commanded by a colonel. A separate regiment is one not assigned to a division or brigade. The British use the **brigade**, a grouping several independent infantry battalions, as the equivalent tactical unit.

Regiment (British Army).—*a.* Used to designate a combatant arm, such as the Royal Regiment of Artillery and the Royal Tank Regiment.

b. Battalion-size units of the cavalry; some battalion-size tank units in the Royal Armoured Corps; and "field regiments" in the Royal Regiment of Artillery, the size of which varies according to type (field, medium, anti-aircraft, ant-tank, etc.).

c. The parent organization for a number of infantry battalions which have never trained nor fought together but which have a historic name in common. The battalions have a regimental training depot in common, but the parent organization has no tactical functions.

d. On occasion, in order to obtain brevity, a battalion of the Royal Tank Regiment (RTR or R Tanks) may be referred to, for example, as 6 RTR or 6 R Tanks (i.e., the 6th Battalion of the Royal Tank Regiment).

Regimental aid post.—*British term: see* **Aid station**.

Regimental reserve area.—An area in which the regimental reserve is usually disposed for defense along and behind the regimental reserve line. *British equivalent:* **Brigade reserve area**.

Regimental reserve line.—A line designated to coordinate the locations and actions of the regimental reserves in a battle position. For example, it may form the line of departure for all planned counterattacks and usually marks the forward limit of artillery position areas for defense of the main line of resistance. *British equivalent:* **Brigade reserve position**.

Regimental S-1.—Regimental adjutant.

Regimental S-2.—Regimental intelligence officer.

Regimental S-3.—Regimental plans and training officer.

Regimental S-4.—Regimental supply officer.

Regimental sector.—*See* Sector. For example, the area defended by a regiment, including the area in front of the MLR to the limit of fire of direct support artillery, and the rear areas of the battle position to include the regimental supply installations.

Regimental train.—Those vehicles that transport a regiment's ammunition, kitchen equipment, baggage, engineers' equipment, and maintenance supplies, and the regiment's medical detachment. A regimental train differs from a company transport, which is composed of vehicles primarily of tactical importance.

Register.—To adjust fire on a target; determine accurate firing data for a target by firing trial shots in preparation for action.

Registration.—An adjustment on a selected point to determine data for use in preparation of fire. *British equivalent:* Same *or* **trial shoot**. (*Also British term: see* **Check concentrations** *and* **Fire for adjustment**)

Registration target.—See **Auxiliary target**.

Regular Army.—A permanent army maintained in peace as well as in war; standing army; one of the major components of the Army of the United States. *British equivalent:* Same.

Regular Army Reserve.—A reserve component of the Regular Army, consisting of honorably discharged members available on call for military service.

Regulating point.—An easily recognizable point where an incoming motor transport column is separated into detachments for entrucking or detrucking purposes. *British equivalent:* Same.

Regulating station.—*a.* A traffic control agency established on lines of communication and through which movements are directed and controlled by the commander of the theater of operations.

b. British: The point on the line(s) of communication at which a transporting agency is given further directions as to destinations of transport. See also **Sector control.**

Regulating unit.—Those smaller units within march units which change gaits as a unit at the command or signal of their respective commanders.

Regulation.—*a.* Any of the official policies and rules for governing and training any branch of the Army.

b. Authorized; according to, or required by, regulation.

Regulation officer.—The officer in charge of a regulating station.

Rehabilitation.—Restoring to former standing or good condition; as, the rehabilitation of a wounded soldier.

Reinforcements.—Troops used to augment the strength of another body of troops, especially for combat purposes. *British equivalent:* Same (*but see also* **Replacement**).

Relative rank.—*a.* The comparative rank or position of authority among officers holding the same grade. *b.* The corresponding rank in another service. An admiral and a general have the same relative rank.

Relay point.—A station or installation at which supplies are transferred from one carrier to another, or at which radio messages are picked up and rebroadcast for forwarding to another relay point or to their final destination; relay station.

Relief.—*a.* Variations in height of the earth's surface.

b. Lines or markings on a map that indicate the various altitudes of points on the earth's surface.

c. Release from a particular duty or assignment. *d.* A replacement of a unit by another unit.

e. A change of soldiers on duty, particularly interior guard duty.

f. A person or unit that takes over or gives aid. The main who takes a sentinel's place is called his relief.

g. British: The length of time that men have to work before being relieved, or a number of men who work, or are on duty, for a given length of time.

Relief map.—A map that shows the different heights of a surface, by using shading, colors, solid materials, etc.

Relocation.—*a.* Determining that range and azimuth from station to a target when the range and azimuth from another station to the target are known. *b.* Determining the range and azimuth of a future position of a moving target.

Relocation clock.—A circular diagram used in fire adjustment to show accurately the positions of a moving target and the deviations of shots as reported by observers.

Remount.—*a.* A horse furnished to replace a mount that has been killed or disabled.

    b. To furnish such a horse.

Rendezvous.—*a.* An appointed meeting place. *British equivalent:* Same.

    b. In bivouacking, the point at which a person assigning bivouac areas meets unit representatives.

    c. British: A place at which **third line transport** is met by guides and directed to **refilling points**. As applied to a **supply column**, the place to which the loaded echelon is moved preparatory to proceeding to **meeting points**.

    d. British: In combined operations, the place of assembly for complete units, to which sub-units proceed from **forming-up places**.

Reorganization.—The restoration of order in a unit after combat. Reorganization includes replacing casualties, reassigning men if necessary, replenishing the ammunition supply, etc.

Repatriate.—An individual who is restored or returned to his own country or citizenship. *British equivalent:* Same.

Replacement.—An individual available for assignment. *British equivalent:* **Reinforcement.**

Repairs, base.—*British term:* see **Maintenance, fourth echelon.**

Repairs, first-line.—*British term:* see **Maintenance, first echelon.**

Repairs, second-line.—*British term:* see **Maintenance, second echelon.**

Repairs, third-line.—*British term:* see **Maintenance, third echelon.**

Replacement of assemblies.—*British term:* see **Unit replacement.**

Report centre.—*British term:* A prearranged position to which reports intended for a commander must be sent.

Reporting line.—A telephone connection between an observation post and a filter room or plotting room.

Report (or reporting) line.—*British term*: A line having not tactical significance, such as a lateral road or railway, on which units or formations report but do not halt. See **Phase line.**

Requirements.—The computed needs for a military force embracing all supplies necessary for its equipment, maintenance, and operation for a given period. They are classified as individual, organizational, initial, maintenance, and reserve. *British equivalent*: Same.

Requirements, individual (initial) (maintenance) (organizational) (reserve).—See **Individual (initial) (maintenance) (organizational) (reserve) requirements.**

Requisition.—*a.* A request for supplies, usually on a form furnished for the purpose. *British equivalent*: **Indent.**

b. Also used to signify the purchase by demand of supplies in hostile occupied territory. *British equivalent*: Same.

Resection.—*a.* The determination of an observer's own position by means of observations taken on points whose positions are known.

b. A method of locating on a map or chart by means of lines drawn from other points located on the map or chart. There are three methods of resection: **back-azimuth, Bessel method**, and **tracing paper method.**

Reserve.—*a.* A fraction of a command held initially under the control of the commander to influence future action. *British equivalent*: Same.

b. The largest element of an advance guard. *British equivalent*: **Main guard.**

c. Soldiers or sailors not in active service, but available for call.

d. British: see **Support (noun).**

Reserve (or rear) battle position.—Any battle position in rear of the main battle position, that has been reconnoitered and staked out, and generally partially organized, for use in case the troops are driven out of the main battle position.

Reserved transportation area.—*British term*: An area surrounding a transportation installation (i.e., canal, railway, docks, etc.), reserve fro the working of the transportation agency alone.

Reserve requirements.—Those supplies necessary to meet emergency situations incident to a campaign. *British equivalent*: **Reserve supplies.**

Reserves (supply).—Supplies accumulated in excess of immediate needs for the purpose of insuring continuity of an adequate supply; also designated as "reserve supplies." *British equivalent:* Same.

Battle reserves are supplies accumulated by the army, detached corps, or detached division in the vicinity of the battlefield in addition to individual and unit reserves.

Individual reserves are those carried on the soldier, animal, or vehicle for his or its individual use in an emergency.

Unit reserves are prescribed quantities of supplies carried as a reserve by a unit.

Reserve supplies.—*British term: see* **Credit** and **Reserve requirements**.

Restricted.—A classification of a military document that is for official use only, that is denied the general public, or that has its circulation limited for reasons of administrative privacy. Restricted items are allowed a wider distribution than those classified as *confidential* or secret.

Retire.—To withdraw or fall back from a position according to a plan, and without pressure from the enemy. *British equivalent:* **Withdraw**.

Retirement.—A retrograde movement of the main forces which, while contact with the enemy is not an essential condition, is generally made for the purpose of regaining initiative and freedom of action by a complete disengagement. A movement made to forestall a decisive engagement, to attract the enemy in a desired direction, or to gain time for the reorganization of the forces preparatory to renewed efforts against the enemy. *British equivalent:* **Withdrawal.**

Retreat.—*a.* An involuntary retrograde movement forced on a command as a result of an unsuccessful operation or combat. The act of retreating. To retire from any position or place. To withdraw. *British equivalent:* Same *or* **withdrawal**.

b. A flag lowering ceremony held at sunset at a military post.

c. A bugle call sounded, with or without beat of drums, at the beginning of the flag lowering ceremony at sunset.

Retrograde defensive.—A defensive withdrawal in which the decisive battle is avoided until preparations are complete for taking the offensive with reasonable chances of success. It delays the enemy, draws him farther from his major supply bases, and elongates his lines of communication.

Retrograde movement.—*A* movement to the rear. *British equivalent:*

Same.

Return.—An official report made to higher authority, especially in accounting for troops, property, or supplies.

Reveille.—A bugle call marking the rising hour at a post or camp.

Reverse slope.—A slope which descends away from the enemy and forms the masked or sheltered side of a covering ridge. The rear slope of a position on elevated terrain. *British equivalent*: Same.

Revet.—To face a wall, embankment, or trench with masonry, sandbags, or other material to add strength and support.

Revetment.—A retaining wall or facing constructed from sandbags, boards, or brush which hols earth slopes at steeper angles that they normally would retain without caving or sliding.

Reviewing authority.—A commander to whom the record of a court-martial action is submitted for review and approval; also known as ***approving authority***. The exercise of this authority is a function of command and not of rank. The reviewing authority is also the ***appointing authority***.

Ricochet burst.—A burst of a high explosive shell in the air after the projectile has hit and bounced. A ricochet burst is used effectively against enemy personnel, instead of the common air burst secured by a time fuze before the projectile strikes.

Ricochet fire.—Fire at a low angle of elevation, with the burst occurring in the air close above the ground after the projectile has hit and bounced.

Rifle.—*a.* Any firearm that has spiral grooves or rifling in the bore designed to give a spin to the projectile for greater accuracy of fire and longer ranges.

   b. A shoulder weapon with spiral grooves cut in the bore. A rifle is one of the important kinds of small arms.

   c. To cut spiral grooves in the bore of a gun in order to give a spin to the projectile so that it will have greater accuracy of fire and longer range.

Rifle grenade.—A grenade, or small bomb, designed to be fired from a rifle using a special device called a ***launcher***, attached to the muzzle of the gun.

Rifleman.—A soldier armed with a rifle as his principal weapon.

Rifling.—Spiral grooves in the bore of a rifle designed to give a spin to the projectile for greater accuracy and carrying power. Rifling includes both the ***grooves*** and the ridges between, called ***lands***.

Right (left) bank of stream.—The bank which is on the light (left) of the observer when facing downstream. *British equivalent*: Same.

Right (left) flank.—The entire right (left) side of a command from the leading element to the rearmost element as it faces the enemy.

Rip cord.—*a.* A control cord that releases a parachute from its container. A parachute jumper pulls the rip cord when he wants the parachute to open.

b. A control cord that tears a piece from the envelope of a balloon or nonrigid airship in order to deflated the balloon rapidly.

Road block.—A barrier to block or limit the movement of hostile vehicles along a road. *British equivalent*: Same.

Road crater.—A hole blown in the road at points which cannot be easily detoured. *British equivalent*: Same.

Road discipline.—The orderly, systematic movement of troops, vehicles, and mounts using a road. Road discipline prevents confusion and delay.

Roadhead maintenance area.—*British term*: An area in the vicinity of roadheads in which are situated advance depots and dumps for the maintenance of forward formations, served by these roadheads, and of local units.

Road space.—The distance from head to tail of a column when it is in prescribed formation on a road. *British equivalent*: Same.

Road time.—The total time a column or a march unit requires to clear a given section of a road.

"Roger."—A world used to indicate that a radiotelephone message has bee received. It is the equivalent of the letter "R", standing for "Received."

Roll call.—A calling off a list of names of the members of an organization.

Rolling barrage.—Artillery fire on successive lines, advancing according to a time schedule and closely followed by assaulting infantry elements. *British equivalent*: **Moving barrage**.

Rolling reserve.—Reserve supplies held close to troop units. When equipment is available, these supplies are kept in railroad cars or in trucks ready for immediate use.

Roll up.—*a.* To attack the enemy's flanks, pushing them toward the center.

b. To attack one or both of the flanks made by a break-through of the enemy front, so as to widen the gap.

Rope ferry.—A set of ropes strung over a stream or defile, over which equipment is moved from one bank to the other. The equipment is rigged on the ropes and pulled across the stream or defile by a tow-line.

Roster.—*a.* A list of personnel.

b. A list of officers and men available for a specific duty with a record of duty each has performed.

Rotating band.—A soft metal band around a projectile near its base. The rotating band has slightly greater diameter that the raised part of the rifling of the bore so that, as the projectile spins through the bore, the rifling cuts into the rotating band. The rotating band makes the projectile fit tightly in the bore by centering the projectile, thus preventing the escape of gas, and giving the projectile its spin.

Round.—*a.* All parts that make up the ammunition necessary in firing one shot. A round consists mainly of a primer, a propelling charge, and a projectile. In fixed ammunition, these three parts of a round are held together with a shell case. In small-arms ammunition, the projectile is called a bullet and a complete round is called a cartridge.

b. One shot fired be each man or each gun of a unit.

Route march.—The advance in column on roads.

Route of communication.—The road net available for tactical maneuver or supply. Routes of communication include rail facilities, navigable waters, and airplane landing facilities. *British equivalent:* Same.

Route reconnaissance.—A careful survey of a route for military purposes, often by aircraft.

Routine message.—A message requiring no special precedence. *British equivalent:* Same.

Routine order.—An order covering matters not concerned with, or affected by, operations in the field. Routine orders include general and special orders, court-martial orders, bulletins, circulars, or memoranda. *British equivalent:* Same.

Roving gun (field artillery).—An artillery piece withdrawn from its regular position and posted in a temporary position for the execution of a specific mission upon the conclusion of which it rejoins its battery. *British equivalent:* Same.

Royal Army Service Corps.—*British term:* The branch of the Army responsible for the storage and issue of supplies, for certain phases of their transportation, including the vehicles assigned for that purpose, and for the administration of barracks and quarters. *Abbrev.* R.A.S.C.

Royal Army Ordnance Corps.—*British term*: The branch of the Army responsible for the procurement and issue of ordnance stores, but all of these stores are eventually to be maintained and repaired by the Royal Electrical and Mechanical Engineers.

Rucksack.—A canvas or leather bag with a shoulder harness, used for carrying clothes and equipment on the back. It is a type of knapsack used by some mountain and ski troops.

Runaway gun.—An automatic weapon that continues firing after the trigger is released. A runaway gun is caused by a defect in some part of its mechanism.

Runner.—A foot messenger. *British equivalent*: Same.

Running fight.—A battle which continues while one side retreats and the other pursues.

Rush.—*a.* A quick short run of foot troops, made in approaches under fire.

> *b.* to make a quick short run of this sort.

S-1, S-2, S-3, S-4.—See ***General staff.***

Sabotage.—The destruction of, or injury to, property by enemy agents or sympathizers in an effort to stop or slow down a nation's war effort.

Saboteur.—An enemy agent or sympathizer who destroys or injures property in an effort to stop or slow down a nation's war effort.

Safety pin.—A pin that fits into the mechanism of a fuze and makes it impossible to set off the fuze accidentally.

Salient.—A portion of a battle line or fortification which extends sharply to the front of the general line. *British equivalent:* Same.

Sally.—See ***Sortie.***

Salvage.—The collection of abandoned, captured, or unserviceable property with a view to its utilization or repair. Property so collected. To recover or save. *British equivalent:* Same.

Salvo.—*a.* A group of shots fired at the same time by a battery, one round each gun.

> *b.* A group of bombs dropped from an airplane at the same time.
>
> *c.* A series of shots fired by a battery. Each gun fires its round in

turn after a given interval

Sam Browne belt.—An officer's leather belt that has a supporting strap worn across the right shoulder.

Sanitation.—The use or application of sanitary measures. *British equivalent:* Same.

Sap.—*a.* A trench that is extended by digging away earth at one from within the trench itself. The earth is usually thrown up to serve as a parapet on the exposed end or flank.

b. To undermine by digging; approach by digging trenches.

Sapper.—A member of an engineer unit trained for digging trenches, tunnels, and underground fortifications.

Scabbard.—A sheath or case; holder for a sword, bayonet, rifle, carbine, or other weapon.

Satellite field.—An airdrome that is completely equipped for all types of servicing or repair of aircraft, but is connected for such operations with a fully equipped airdrome. Also known as an *auxiliary airdrome*.

Schedule fires.—See *Prearranged fires*.

*Scheme of command.*A plan for the control of all elements of a command during a military operation including provision for communication, observation, and the location of the command post.

Scheme of fire.—The part of a military plan that deals with the arrangements for the locations and missions of the different weapons and for the control of their fire in either defense or attack.

Scheme of maneuver (plan of attack).—The commander's plan for employing subordinate units to accomplish a mission. For example, in the attack it usually distributes the troops for two general missions: A *holding attack*, usually frontal; and the *main effort*, which may seek penetration or envelopment. *British equivalent:* *Plan of attack*.

Scout.—A man specially trained in shooting, in using ground and cover, in observing, and in reporting the results of observation; a man who gathers information in the field; to reconnoiter a region or country to obtain information of the enemy or for any other military purpose; to act as a scout. *British equivalent:* Same.

Scout airplane.—An airplane engaged in an aerial search, or used to reconnoiter enemy positions and territory by direct observation or by photography.

Scout car.—An armed and armored motor vehicle used primarily for reconnaissance. *British equivalent*: Same (but without a cover).

Screen.—To prevent hostile ground reconnaissance or observation. The body of troops used to screen a command.

Screening smoke.—A chemical agent used to blind hostile observation. *British equivalent*: Same.

Seacoast artillery (Coast Artillery Corps).—All artillery weapons used primarily for fire upon hostile naval vessels. It includes both fixed and mobile armament. *British equivalent*: **Coast defence artillery**.

Seaplane.—An airplane equipped with floats that support it on the water so that it can come down or take off at sea. Airplanes with boat-shaped hulls which support them in the water are more properly called *flying boats*.

Searching fire.—Fire distributed in depth by successive changes in the elevation of the gun. It is opposite of ***traversing fire***, which is distributed in width by moving the gun gradually to the right or left.

Search patrol method.—One of three methods of using fighter aviation in air defense. With this method fighter aircraft are sent out on a systematic search of a given area for enemy aircraft. Other methods are the ***air alert method*** and the ***ground alert method***.

Secondary artillery (Coast Artillery Corps).—Seacoast artillery weapons of less than 12-inch caliber. *British equivalent*: **Heavy coast defence guns** (approximately).

Secondary attack.—See **Holding attack**.

Second line transport.—*British term*: see **Field Transport, R.A.S.C.** and **Transport (British)**.

Secret.—A classification given to an official document when the disclosure of its contents might endanger national security, cause serious injury to government activity, or be of great advantage to the enemy. Secret documents are available only to those whose duties require the information contained in them. They are more closely guarded than those marked ***restricted*** or ***confidential***.

Secret text (or secret language).—The text of a message which, on its face, conveys no intelligible meaning in any spoken language. The secret text of a message constitutes a cryptogram. *British equivalent*: **Message in cipher or code**.

Sector.—The defense area designated by boundaries within which a unit operates on the defense. (Ordinarily used when referring to regiments and higher units.) One of the subdivisions of a coastal frontier. See **Defense area**.

Sector of fire.—A section of terrain designated by boundaries assigned to a unit or to a weapon to cover by fire. For example, the sector begins at the position or emplacement of the weapon or unit and is limited on its sides by prescribed boundaries, and at its extremity by the limit of range of the weapon or weapons or by observation at the firing position(s). The boundaries may be defined by the designation of points or, in the case of weapons, by an arc of fire. Compare **Target area**. *British equivalent*: **Arc of fire**.

Secondary target area.—Any target area(s) assigned as secondary fire mission(s) to a weapon or unit to be engaged when not required to fire on its primary target. Priority of engagement may be assigned numerically in order of importance to several secondary target areas assigned to one unit or weapon. Compare **Primary target area**.

Second-line trench.—*British term*: see **Support trench**.

Section.—*a*. A subdivision of an office or organization; especially, a major subdivision of the staff of units smaller than a division.

b. A tactical unit of the Army. A section is smaller than a platoon and larger than a squad. In some organizations the section, rather than the squad, is the basic tactical unit.

c. British: (i) For cavalry (and artillery): the four front rank men and their covers; (ii) Artillery: one of the divisions of a battery with its complement of men, horses, and wagons; (iii) Engineers: the subdivision of a company; (iv) Infantry: the subdivision of a platoon; (v) Tanks: five tanks, the subdivision of a company.

Section II.—A discharge from military service for disability, so called because it is authorized by Section II, Army Regulations 615-360. The discharge may be honorable or without honor, depending on the facts in the particular case.

Section VIII.—A discharge from military service for unfitness, so called because it is authorized by Section VIII, Army Regulations 615-360. The discharge may be honorable or without honor, depending on the facts in the particular case.

Sector.—A clearly defined area which a given unit protects or covers with fire; part of a front held by a unit. Sometimes only the sections held by regiments or larger units are called sectors; those held by battalions or smaller units may be called areas. *British equivalent:* Same.

Sector control.—*British term*: see **Regulating station.**

Secure.—Same as **Seize.**

Security.—All measures taken by a command to protect itself from observation, annoyance, or surprise attack by the enemy; and to obtain for itself the necessary freedom of action. The protection resulting from such measures. *British equivalent:* **Protection** (also a classification of protected papers, equivalent to U.S. "Restricted").

Security detachment.—Any unit disposed to protect another unit against surprise or interference by the enemy. *British equivalent:* **Covering detachment.**

Security mission.—A task or duty of protecting and screening friendly forces against enemy attack or observation.

Seize (or secure).—To gain physical possession of, with or without force. *British equivalent:* Same.

Selectee.—A person inducted into military service under the provisions of the Selective Service Act; draftee.

Selective Service.—The selection of persons from the total manpower of a country for compulsory military service. Also called **draft.**

Self-loading.—of guns, using the force of recoil or of gas pressure to throw out a spent shell and to place a fresh one in the chamber ready to fire.

Self-propelled artillery.—Artillery mounted on motor vehicles on which it is carried, and from which it is fired.

Self-propelled gun.—An artillery weapon carried on a self-propelled mount. **Tank destroyers** are self-propelled guns.

Semaphore.—*a.* A method of signaling with flags, in which the signalman holds a flag in each hand and signals by holding them in different positions.

    b. To signal by such a system.

Semiautomatic.—Partly self-acting; as applied to a gun, self-loading but not self-firing. A semiautomatic gun throws out the used shell and puts in a new one by its own recoil action or gas pressure, but a separate pull of the trigger is needed for each shot. Semiautomatic means partly self-acting, as distinguished from **automatic**, which means completely self-acting

Semiautomatic fire.—Fire delivered from a self-loading weapon, with a trigger pul for each shot fired. It differs from **automatic fire**, in which the gun continues to fire until the pressure on the trigger is released.

Semiautomatic weapon.—A gun that is automatically loaded and prepared for firing after each shot, but that fires only one shot with each trigger pull.

Semifixed ammunition.—Ammunition in which the size of the propelling charge can be adjusted to correct for range, and which is loaded into the cannon as a unit.

Seniority.—The precedence in rank that one soldier has over another by reason of date of commission or appointment, length of active service, etc. When two persons are appointed to the same rank on the same day, the person with the longest active service has seniority.

Sensing.—*a.* Observing the direction of the striking or bursting point of a projectile from the target, and reporting whether the shot is a hit, over or short, right or left, lost or doubtful. Sensing does not include accurate measurement of distances and angles.

b. Determining with a directional antenna from what direction a radio wave is approaching a receiving set.

Sentry.—A soldier assigned to duty as a member of a guard, to keep watch, maintain order, protect persons or places against surprise or warn of enemy attack; sentinel.

Sentry squad.—A squad posted for security and information with a single or double sentinel in observation, the remaining men resting nearby and furnishing the relief for the sentinels. An outguard of one squad. *British equivalent:* **Sentry post.**

Separate battalion.—A battalion that does not form part of a regiment, and that operates as an independent unit it the field. It is an administrative as well as a tactical unit.

Separate battery.—An artillery battery that does not form part of a regiment or battalion, and that operates as an independent unit in the field.

Separate-loading ammunition.—Ammunition in which the projectile, propelling change, and primer are not held together in a shell case, as in *fixed ammunition*, but are loaded into a gun separately.

Sergeant.—*a.* A noncommissioned officer who holds rank next above that of a corporal and next below that of a staff sergeant or technician third grade; enlisted man of the fourth grade.

b. A title of address for an enlisted man of any of the first four grades.

Sergeant major.—*a.* The chief administrative clerk in a battalion or higher unit, usually holding the rank of technical sergeant or of master sergeant.

b. The highest noncommissioned officer in the U.S. Marine Corps, equivalent to chief petty officer in the Navy and to master sergeant in the Army.

Sergeant of the guard.—A title given to the senior noncommissioned officer of an interior guard, no matter what his grade may be.

Serial.—One or more march units, preferably with the same march characteristics placed under one commander for march purposes.

Service.—Any branch of the Army mainly concerned with administration, supply, transportation, or medical care; noncombatant branch; one of the subdivisions of the Army Service Forces. A combatant branch is called an **arm**. *British equivalent:* Same.

Service ceiling.—A height above sea level, under normal conditions, at which an airplane is unable to climb faster than 100 feet a minute.

Service command.—An organization serving as a field agency of the Army Service Forces in a specified area. Its mission is to provide necessary services, including administrative, financial, legal, statistical, medical, welfare, etc., for elements of the Army; to construct facilities and provide fixed communication services; and to procure, store, maintain, and distribute supplies and equipment both within and without the limits of the service command. Formerly called **corps area.**

Service of a piece.—The operation and maintenance of a gun or other equipment by its crew.

Service park.—A place in the forward area where the organic maintenance elements and attached medical elements assemble when combat units enter combat.

Service pilot.—*a.* The rating given to a member of the U.S. Army Air Forces who is qualified to pilot aircraft other than combat airplanes.

b. A person who has this rating.

Service train.—Formerly, the train of any unit serving the division as a whole rather than any particular unit. See **Train**. *British equivalent: **Supply units**.*

Service units (or elements).—Those organizations provided for by Tables of Organization within larger units whose functions are to provide for the supply, transportation, communication, evacuation, maintenance, construction, and police of the larger unit as a whole. *British equivalent:* **Supply units — "The Services"** (provided by War Establishments).

Shell.—*a.* A hollow projectile designed to be fired from a weapon. It may contain an explosive, or a chemical or other filler. It may be fixed or separate-loading.

   b. A shotgun cartridge.

   c. To bombard; to fire a number of projectiles at a target.

Shell crater.—A bowl-shaped hole left in the earth by the explosion of a projectile; shell hole.

Shell fragments.—Jagged pieces of metal scattered by the burst of a shell. High explosive shells, with the inside of the case scored so as to give high fragmentation, had largely replaced shrapnel during World War II for purposes requiring a burst in the air.

Shellproof.—Capable of resisting bombs or projectiles.

Shelter.—Any form of concealment from view, of protection against the elements or the fire of weapons. That which covers or defends. A screen. Protection. To afford or provide shelter. To screen or cover from notice. *British equivalent:* Same *or* **cover**.

Shelter half.—One half of a small two-man tent. Each man carries his half as part of his field equipment.

Shelter, heavy shellproof.—A shelter which protects against continuous bombardment by at least 8-inch shells. *British equivalent:* None.

Shelter, light.—A shelter which protects against direct hits, and in some cases against a continued bombardment, by 3-inch shells. *British equivalent:* None.

Shelter, light shellproof.—A shelter which protects against continuous bombardment by all shells up to and including the 6-inch. *British equivalent:* None.

Shelter, splinter-proof.—A shelter which protects against rifle and machine-gun fire, against splinters of high-explosive shell, and grenades, but not against direct his by 3-inch shells. *British equivalent:* Same.

Shelter tent.A small tent capable of providing shelter for two men. It is made in two halves, one of which is carried by each man as part of his field equipment.

Shelter trenches.—Hasty trenches constructed to provide shelter from fire and to permit riflemen to fire in a prone position. See

*Foxhole. British equivalent: **Slit trenches.***

Shock action.—A method of attack by mobile units in which the suddenness, violence, and mass weight of the first impact produces the main effect. Tank attacks usually rely on shock action.

*Shock army.*See *Army.*

Shock troops.—Troops especially organized, trained, and equipped for assault and hand-to-hand fighting.

Short.—A shot which strikes or bursts between the gun and the target; shot fired without sufficient range to reach the target.

Short.—A shot which strikes or bursts between the gun and the target; a shot fired without sufficient range to reach the target. Shorts and overs are observed in *sensing.*

Shot.—*a.* A projectile which is solid, as contrasted to a *shell*, which is hollow and contains explosive, or a chemical or other filler.
   b. Small balls or slugs used in shotgun shell, canisters, and some other types of ammunition. *c.* A single discharge of a firearm of any type.

Shouldering process.—Attacking an enemy position by alternating fire and movement, so that the moving elements are protected by fire delivered by the stationary one.

Shoulder weapon.—Any small-arms weapon designed to be fired while held in the hands, with its but braced against the shoulder. The rifle, carbine, shotgun, and automatic rifle are shoulder weapons.

Shrapnel.—An artillery projectile which contains small lead balls that are released by a powder change in the base, set off by a time fuze. Although it was still in service during World War II, shrapnel had been replaced almost entirely by high explosive shells. Wounds called shrapnel wounds usually were due to shell fragments rather than to shrapnel.

Shroud.—One of the ropes attached to the edge of a parachute canopy, used to support the man or object being dropped.

Side arms.—Weapons that are worn at the side or in the belt when not in use. The sword, bayonet, automatic pistol, revolver, etc. are side arms.

Side spray.—Fragments of a bursting shell that are thrown out to the side of the line of flight in contrast with *nose spray,* which are fragments of a bursting shell that are thrown out to the front, and *base spray,* which are fragments of a bursting shell that are thrown to the rear.

Siege warfare.—A type of warfare that develops when an attacking force wholly or partly surrounds a strongly defended position and attempts to wear it down by blockade, bombardment, and small, systematic advances.

Signal.—*British term*: see **Message**.

Signal centre.—*British term*: A special **signal office**, at a point where means of communication with two or more headquarters may be concentrated, or at a sight to which a headquarters may move and establish a signal centre. See **Message center**.

Signal communication.—All methods and means employed to transmit messages and telephone conversations. *British equivalent*: Same.

Signal communication along the centre line.—*British term*: see **Axis of signal communication**.

Signal communication along the main axis of advance.—*British term*: *see* Axis of signal communication.

Signal operation instructions.—A type of combat orders issued for the technical control and coordination of signal agencies throughout the command.

Signal flare.—Any pyrotechnic signal, fire from the ground or from aircraft, that has distinctive color or characteristics which give it meaning in an established code system.

Signal office.—*British term*: An office established at every headquarters provided with intercommunication personnel to deal with signal traffic in a coordinated and systematic manner. See also: **Signal centre** and **Message center**.

Signal pistol.—A pistol from which pyrotechnics, especially flares, are fired. Also called **pyrotechnic pistol**.

Signal rocket.—A rocket that gives off some characteristic color or display which has a meaning according to an established code. It usually comes from a **signal pistol** or a **ground signal projector**.

Signal security.—The security of friendly signal communication message traffic against the availability and intelligibility of that traffic to enemy or other intelligence agencies. *British equivalent*: Same.

Signal unit.—*British term*: see **Agency of signal communication**.

Single action.—A method of fire in a revolver and in old-style rifles and shotguns that necessitates cocking the hammer by hand before firing, in contrast to **double action**, in which a single pull of the trigger both cocks and fires the weapon.

Site defilade.—See **Position defilade.**

Situation.—All the conditions and circumstances, taken as a whole, which affect a command at any given time, and on which its plans must be based. They include such items as the positions, strength, armament, etc., of the opposing forces and any supporting troops, considerations of time and space, the weather, terrain, etc., and the mission to be accomplished. A consideration of these conditions and the possible courses of action open, followed by a decision, constitute the estimate of the situation. *British equivalent*: Same (a consideration of these conditions, and the possible courses of action, followed by an intention, constitutes the appreciation of the situation).

Situation map.—A map showing the tactical or administrative situation at a particular time, usually for use as a graphic aid in carrying on the work of a staff section or as an annex to staff reports. *British equivalent*: **Situation (or battle) map.**

Skirmish.—*a*. A relatively light engagement between small bodies of scattered troops.

　　b. To take part in a skirmish.

Skirmishers.—Soldiers, dismounted, deployed in line and in extended order in drill or attack. *British equivalent*: None.

Skirmish line.—A line of troops in extended order during a tactical exercise or attack.

Ski troops.—Troops, usually Infantry, equipped with and trained to maneuver on skis. *British equivalent*: Same.

Sketch map.—*British term*: see **Operation map.**

Sling arms.—To place a rifle or carbine with its strap over the shoulder in a position for marching, or for review under certain conditions.

Slit trench.—A very narrow trench used for protection against shell fire and passage of tanks, especially in massing troops close to the front. (*Also British term*: see **Fox hole** and **shelter trenches**)

Small arms—*a*. Firearms of small caliber, including pistols, rifles, machine guns, and shotguns. The maximum caliber for small arms is set variously in different branches of the service, commonly either .60 calibers or one inch.

b. Weapons which may be carried by individuals.

Small scale maps.—Maps of a scale varying from 1:1,000,000 to 1:7,000,000, intended for the general planning and strategical studies of the commanders of large units. *British equivalent*: Same (*British usage also includes* **Intermediate-scale maps**).

Smokeless powder.—A propellant explosive from which there in little or not visible smoke on firing.

Smoke pot.—A small smoke generator used to produce a screen or blanket of smoke.

Smoke projectile.—Any projectile containing a smoke-producing chemical agent that is released on impact or burst. Also called a **smoke shell**.

Smoke screen.—Curtain of smoke employed for masking either friendly or hostile activities or installations.

Smooth-bore.—Having a bore that is smooth and without rifling. Shotguns and mortars are commonly smooth-bore.

Sniper.—A soldier, usually and expert shot, detailed to fire at and pick off individuals of the enemy.

Soft spot tactics.—Tactics in which tanks or other assault elements strike through spots of least resistance, bypassing strongly defended localities and leaving them to be reduced later by other troops attacking from the flank or rear.

Soldier.—Any member of the Army. Sometimes the word is used to mean an enlisted man as contrasted to an officer. *British equivalent*: Same.

Sortie.—*a.* A sudden attack made from a defensive position. In this meaning also called **sally**.

b. A single round trip of an aircraft on a tactical mission.

Sound ranging.—A method of locating the source of a sound, such as that of a gun report or a shell burst, by calculations based on the intervals between the reception of the sound at various previously oriented microphone stations. Sound location of aircraft by means of a sound locator is also sometimes called sound ranging.

Spearhead.—*a.* The leading assault troops in an attack.

b. To hold the leading position in an assault or rapidly moving attack.

Special court-martial.—A court martial made up of at least three officers appointed by the commanding officer of a station, regiment, or higher unit, to try enlisted men of the grade of master sergeant or lower for noncapital crimes. *British equivalent*: **District court-martial.**

Special orders.—Instructions issued in order form that apply only to individuals or small groups of individuals, rather than to a command as a whole. Special orders usually cover such maters as assignment, reassignment, transfer, promotion, separation, appointment of boards, etc. Orders applying to a whole command are called **general orders.**

Special force.—*British term*: see **Task force.**

Special staff.—*A* staff group, subordinate to the general staff of a unit, whose duty it is to assist the commander in the exercise of his tactical, administrative, technical, and supply functions. It includes the heads of the administrative, technical, and supply services, and certain technical sPecialists. In divisions and higher units the general and special staffs are separate, but in lower units they partly merge into each other. A special staff officer may also exercise command in his own branch. *British equivalent*: None.

Speed.—The rapidity of movement at any particular instant expressed in miles per hour. (*British usage also for* **Rate of march**.)

"Splashes."—*British term*: see **Identifications.**

Split trail.—A trail divided into two parts, used in certain types of artillery.

Spoil.—Dirt removed in digging trenches and other excavations, lying on the surface of the ground. *British equivalent:* Same.

Sponson.—A hollow enlargement on the side of the hull of a tank, used for storing ammunition, or as a space for radio equipment or guns.

Spotter.—A person who observes the striking or bursting point of a projectile in order to supply information for the adjustment of fire.

Springfield rifle.—United States rifle, .30-caliber, M1903. It is a breech-loading, magazine-fed, bolt-operated weapon.

Squad.—*a*. A group of enlisted men organized as a team; smallest tactical unit, consisting of only as many men as a leader can direct easily in the field. *British equivalent*: **Section.**

 b. British: A small body of men formed for drill or for work.

Squadron.—*a*. The basic administrative and tactical unit of the Army Air Forces. A squadron is larger than a flight, smaller than a group. It is composed of two or more flights. A squadron is the equivalent to a battery in the artillery, a troop in the cavalry, or a company in other branches of the service.

b. British: The fighting unit of the R.A.F. Squadrons are designed army co-operation, bomber, fighter, communication, bomber-transport, according to their nature. The squadron consists of a number of flights each of an aircraft strength varying with the different categories.

c. An administrative and tactical unit of the Army, in the cavalry. It is composed of two or more troops of cavalry. A squadron is equivalent to a battalion in other branches of the Army. *British equivalent:* **Regiment (cavalry)**.

d. *British*: An administrative and tactical unit in the cavalry equivalent to a company in the infantry.

Square division.—An infantry division which has two infantry brigades, each of which has two regiments. It is called a square division because it has four infantry regiments, instead of three, as in a **triangular division**.

Squib.—A small charge of black powder packed around an electric wire and encased in a small tube, one end of which is sealed; flash fuze. It is used to ignite burning-type munitions, such as the gas candle or smoke pot, by means of an electric current.

Stabilized front.—A strong defense line in which the combat elements are distributed over a deep area and in which the flanks are covered or protected by strong barriers. It is usually fortified with trenches.

Stabilized warfare.—Warfare in which there is only limited maneuvering by either side. Trench warfare is a form of stabilized warfare.

Staff.—A group of officers who assist he commander of a battalion or larger unit in carrying out his duties. The staff of a division or higher unit is divided into a general staff and a special staff.

Staff, The.—*British term*: see **General Staff**.

Staff authority.—The authority exercised by a staff officer. (A staff officer, as such, has no authority to command. All responsibility rests with the commanders, in whose name all orders are given.) *British equivalent*: None.

Staff car.—An automobile used by a commander or members of his staff.

Staff sergeant.—A noncommissioned officer in the third grade of the Army, who ranks next above a sergeant and next below a technical sergeant.

Standard.—*a.* A flag carried by mounted or motorized units. **To the standard** is a bugle call sounded as a salute to the color, to the President, the Vice-President, an ex-President, or a foreign chief magistrate. To the standard is usually called **to the color**.

b. A classification given to supplies and equipment that are accepted for general use in the Army.

c. A support, as for the barrel of a mortar.

Standing Patrol.—*British term*: A party of from two men to a troop, or even more, posted a considerable distance in advance of other troops, to watch either the enemy, a route by which he might advance, or a locality in which he might concentrate unseen.

Staging area.—One of a series of areas on the route of march occupied by troops for a long halt. *British equivalent:* Same, also **Rest camp**.

Standing barrage.—*A* stationary artillery or machine-gun barrage laid for defensive purposes in front of an occupied line or Position. Fire on a line, usually placed across a probable avenue of enemy approach, or an exposed sector of the front, in order to prevent passage of enemy troops. *British equivalent:* Same.

Standing operating procedure.—Routine procedure or general order prescribed to be carried out in the absence of orders to the contrary. *British equivalent*: **Standing orders**.

Standing patrol.—*British term*: see **Picket**.

Star shell.—A projectile which contains a chemical that ignited when the projectile bursts. The chemical burns with a brilliant flame and is used to illuminate targets at night.

Start(ing) line.—*British term*: see **Line of departure**.

Starting point.—*British term*: see **Initial point**.

Station.—*a.* A place at which a military organization is stationed, such as a fort or camp. In this meaning, it is also called a **military post**.

b. A position occupied by an individual or unit in the field, such as a clearing station, gun station, or searchlight station.

c. A post of duty of an individual or unit.

d. To post; to place; to assign to a station.

Station designator.—A combination of two or three call letters to identify a radio station.

Station dispensary.—An army medical establishment that provides medical and dental care for military personnel receiving treatment but not needing hospitalization. It serves the men of the post or station on which it is located. An establishment that provides medical and dental care for military personnel not stationed there is called a *general dispensary.*

Sten machine carbine.—A British light-weight machine carbine (or submachine gun), 9-mm caliber. Capable of being manufactured rapidly and at small cost, it was used extensively by air-borne forces, tank crews, reconnaissance units, etc. Also known as the **Sten gun.**

Stereo-pair.—Two vertical aerial photographs taken preferably with an overlap of not less than 60 percent nor more than 75 percent. *British equivalent*: Same.

Stereo-triplet.—Three vertical aerial photographs taken so that the entire area of the center picture is overlapped by the other two.

Sternutator.—An irritant smoke.

Stokes mortar.—See **Trench mortar.** *British equivalent*: Same.

Stores.—*British term*: War material other than **supplies** and are and comprise the following:

    Ordnance stores.—Personal and unit equipment, including armament and small arms, ammunition, explosives, engineer and signal stores, tanks, armoured cars, tractors, carriers and mechanical transport vehicles other than those belonging to the Royal Army Service Corps (R.A.S.C.) units or driven by R.A.S.C. drivers, clothing and necessaries, barrack and camp (movable) equipment, and materials for workshops.

    Engineer stores.—Material and plant, other than unit equipment, required for engineer work of all kinds whether carried out by engineers or other arms. The term embraces permanent line signal stores, but does not include explosives.

    Transportation stores.—Material and plant peculiar to the transportation services, and not provided by other services.

    R.A.S.C. stores.—Mechanical transport vehicles of R.A.S.C. units or driven by R.A.S.C. drivers, spare parts and material and plant for their repair.

    Medical and veterinary stores.—Drugs, dressings, medical and veterinary instruments and appliances.

    Stationary stores.—Stationary, army forms and books, printing and office machinery.

Straddle.—*a.* A group of shots or bombs of which some fall short of, and others pass over, the target, or of which some go to the left of, and others to the right of, the target.

b. To fire a group of shots or drop or group of bombs of which some fall short of, and others pass over, the target, or of which some go to the left of, and others to the right of, the target.

Straggler.—A soldier who has become separated, without authority, from his organization. A motor vehicle that has fallen behind for any reason in an advance. *British equivalent:* Same.

Straggler line.—A line, usually designated by means of well-defined terrain feature such as roads, railroads, or streams, along or in rear of which the military police patrol in order to apprehend stragglers moving to the rear. *British equivalent:* ***Stragglers' posts.***

Straggler post.—The point at which straggling vehicles or personnel are directed to proceed and where they are assembled into groups to be conducted to their units.

Strategic withdrawal.—Moving from a forward defense line to a line further back that offers a better position from which to carry out a combat mission. It is not necessarily a forced retreat, but is usually made for tactical reasons, such as to lure the enemy into a trap, to wear down his combat power, or to maneuver a better position from which to launch an attack.

Strategy.—Making plans and using military forces and equipment for the purpose of gaining and keeping the advantage over the enemy in combat operations. It includes the distribution, transportation, and employment of troops and supplies, and also a study of the combat area, and the disposition and possible lines of action of the enemy. Strategy involves planning on a large scale; tactics involves the operations necessary to carry out these plans.

Stretcher-bearers.—*British term:* see ***Company aid men.***

Striking echelon.—The main attacking elements of a combat unit; a fighting force that engages the enemy in actual combat; striking force.

Striking force.—See ***Striking echelon.***

Striking forces.—Aviation assigned to operate as strong offensive air units for the application of air power. These forces were required to extend the destructive effect of air operations over both land and sea to great distances beyond their operating bases. Tactically they conduct air force operations to gain and maintain control of the air. Compare ***Defense forces*** and ***Support forces.***

Strip mosaic.—A mosaic compiled by assembling one strip of vertical aerial photographs taken on a single flight. *British equivalent*: Same.

Strong point.—Formerly, the defensive area of an infantry company. See **Defense area**. *British equivalent*: **Company sector**.

Subaltern.—*British term*: An officer in the army below the rank of captain.

Subgroupment.—A subdivision of a tactical artillery command, called a groupment, which is formed by the temporary grouping of two or more battalions or larger tactical units that have been assembled from different organizations.

Submachine gun.—A light-weight automatic or semiautomatic gun, designed to be fired from the hip.

Subsector.—Usually one of the subdivisions of a sector. *British equivalent*: Same.

Substitution cipher.—A method of putting a message into code, in which code letters or symbols replace the letters of the plain text, but in which there is no changing or scrambling of the letter order of the words in the message.

Successive attack.—An attack delivered in a series of thrusts in which the various elements of the combat force engage in action, one after the other.

Successive concentration.—The massing of artillery fire on one area after another, beginning with the nearest target. The flexibility of successive concentration permits artillery action to be adapted to support infantry action.

Successive formation.—A movement in which the different elements arrive in their proper places in formation successively.

Summary court-martial.—A court-martial composed of one officer, appointed by the commander of a military station, regiment, or higher unit, to try enlisted men of the grade of staff sergeant or lower for noncapital offenses. *British equivalent*: None.

Superelevation.—The amount that the axis of the bore of a gun must be pointed above the line joining the gun and the target to allow for the curved path of the projectile in flight.

Super-heavy coast defence guns.—*British term*: see **Primary armament (Coast Artillery Corps)**.

Superimposed fire.—*British term*: see **Emergency barrage**.

Supplementary firing position.—A fire position assigned to a unit or weapon to accomplish secondary fire missions other than those to be accomplished from primary or alternate positions. Compare *Primary firing position*; *Alternate firing position*. *British equivalent*: None.

Supplies.—*a.* A general term covering all things necessary for the equipment, maintenance, and operation of a military command, including food, clothing, equipment, arms, ammunition, fuel, forage, and materials and machinery of all kinds.

    b. British: Food, forage, petrol, lubricants for mechanical transport vehicles, fuel, light, disinfectants, and medical comforts.

Supply columns.—*British term*: R.A.S.C. transport columns that carry supplies, mails, engineer and ordnance stores and work between **railheads** and **delivery points**. Each column is divided into two similar echelons which deliver supplies on alternate days.

Supply lines.—*British term*: see **Lines of communication**.

Supply point.—*a.* A general term used to include depots, railheads, dumps, and distributing points. *British equivalent*: **Supply installation**.

    b. British: (i) A point, fixed by divisional headquarters, to which supply column vehicles will be moved. Units will, as ordered by their formations, send their own transport to collect supplies from this point or send guides to lead R.A.S.C. vehicles forward to **delivery points**. (ii) The place at which supplies are transferred from R.A.S.C. transport to unit transport. See also: **Distributing point**.

Supply train.—A motor caravan or train of railway freight cars that transports supplies or equipment for a military organization.

Support (noun).—The second echelon (reserve) of a rifle company (troop) or platoon in attack or defense.

Support (of the advance guard).—The echelon of the advance guard that precedes the advance-guard reserve. The support sends out, and is preceded by, the advance party. *British equivalent*: **Rear section of the van guard**.

Support (of the rear guard).—That part of the rear guard which marches behind the reserve and protects it by observation and combat. *British equivalent*: None.

Support (of the outpost).—The principal echelon of resistance of an outpost. For example, it is usually established along the outpost line of resistance to block sudden enemy attack. Outpost supports furnish outguards to establish security and observation for their positions. A rifle platoon or rifle company usually constitutes a support of an outpost. Where an outpost establishes more than one support, supports are numbered from right to left within the outpost. *British equivalent*: None.

Support forces.—Aviation assigned to provide the necessary air power in support of the operations of ground, naval, or composite forces. While all tactical aviation is trained within its means to provide effective air support for these forces, observation, light and dive bombardment, troop carrier, and photographic aviation are specially trained to furnish air support. In addition to these types of aviation, support forces may include heavy and medium bombardment and fighter aviation. Compare **Striking forces** and **Defense forces.**

Supporting arm.—*British term*: see **Auxiliary arm**.

Supporting distance.—Generally, that distance between two units which can be traveled in the time available in order for one to come to the aid of the other. For small infantry units: That distance between two elements which can be effectively covered by their fire. *British equivalent*: None.

Supporting fire.—Fire delivered by auxiliary weapons for the immediate assistance of a unit during an offensive or defensive action. *British equivalent:* Same.

Supporting distance.—When applied to small infantry units, that distance between two elements which can be effectively covered by the fire of either. *British equivalent*: Unit in support.

Supporting unit.—A unit acting with but not under the direct orders of another unit to which it does not organically belong. *British equivalent: **Unit in support**.*

Support trench.—A fire trench constructed a short distance in the rear of the front-line trenches to shelter the support. *British equivalent: **Second-line trench**.*

Surface line.—A telephone or telegraph line that is laid on the ground hastily during the movement of units in the field. In an organized area, surface lines are replaced by more permanent installations.

Sustained defense.—Defense that is aimed at stopping an enemy attack at the defense line. It differs from a **delaying action**, which attempts to hold off a decisive engagement until the defense forces are in a better position for battle.

Squad defense area.—*See **Defense area**. For example, a squad defense area comprises a section of the platoon defense area assigned to a rifle squad as its task in the all-around defense of the platoon area.

Surveillance.—An active, thorough, and continuous search by observation and reconnaissance of an area or of hostile dispositions.

Sustained defense.—A defense that is aimed at stopping an enemy attack at the defense line. It differs from **delaying action**, which attempts to hold off a decisive engagement until the defense forces are in a better position for battle.

Sweep.—*a.* A swift flight of a formation of combat airplanes over enemy territory.

b. To cover a wide area with gunfire.

Switch position.—A defensive position diagonal to, and connecting, successive defensive positions paralleling the front. *British equivalent:* Same.

Switch trench.—A trench diagonal to, and connecting, successive trenches that are parallel to the front.

Tables of Allowances.—Tables which show the allowance of equipment authorized for posts, camps, an stations. Such equipment in not usually taken with a unit into the field or on change of station.

Table of Basic Allowances.—Tables revised and published to show for each typical administrative unit of the field forces its current authorization of amounts and kinds of basic equipment and supplies, including allotments of armament and ammunition. *British equivalent*: G 1098.

Tables of Equipment.—A list of items of equipment authorized to be taken with an organization on change of station and, under normal conditions, into the field. The tables of equipment do not include expendable items, ammunition allowances, and items of individual clothing and equipment.

Tables of Organization.—Tables published and revised as necessary to show the authorized details of the organization of each typical administrative unit of the field forces. *British equivalent*: **War Establishments**.

Tables of Organization and Equipment.—Tables published and revised as necessary to show the authorized details of the organization of each typical administrative unit of the field forces as well as a list of items of equipment authorized to be taken with an organization on change of station and, under normal conditions, into the field. The tables of organization equipment do not include expendable items, ammunition allowances, and items of individual clothing and equipment.

Tactical groupings.—The balanced grouping of combat units and means within a command to accomplish a tactical mission. It may be accomplished by Tables of Organization, by standard operating procedure within a command, or improvised for a particular operation. *British equivalent:* Same.

Tactical headquarters.—*British term*: see ***Advance command post.***

Tactical obstacles.—Obstacles whose chief purpose is to hold the attacking forces under the effective fire of the defense. *British equivalent:* Same.

Tactics.—The art of handling units in combat; planning and carrying out movements before and during battle, and using combat power on the field of battle. Tactics is also different from ***strategy***, which refers to the broad plans of a nation at war.

Tail of column.—Last element of a column in order of march. *British equivalent:* Same.

Tank.—A full-track armored vehicle carrying machine guns and, usually, a cannon. Tanks are usually classified as light (up to 25 tons), medium (25-40 tons), and heavy (over 40 tons). *British equivalent:* Same (classified as light (considered obsolete by 1942), medium, close-support, cruiser, and infantry tanks, depending on their designed role).

Tank barrier.—A barrier or group of obstacles intended to stop enemy tanks.

Tank defile.—A narrow place, such as a bridge or a road through a mountain pass, that tanks cannot detour. A tank defile is the most effective place to put antitank obstacles.

Tank destroyer.—A self-propelled antitank gun; destroyer.

Tank ditch.—A deep ditch prepared as an obstacle to enemy tanks. It is usually called an ***antitank ditch.***

Tank traps.—Concealed ditches placed in roads, level stretches of ground, or other similar practicable routes of approach so designed that vehicles will fall into them and not be able to escape. *British equivalent:* Same.

Taps.—The last bugle call at night. Taps is blown as a signal that all unauthorized lights are to be put out. Taps is blown also as last honors at military or naval funerals.

Target.—The specific thing at which fire is to be directed.

Target area.—An area assigned to a unit or to a weapon to cover by fire. The limits of the area are prescribed by the commander assigning the area. Firing positions are selected to engage any targets appearing in the area. Compare **Sector of fire.** (For classification of target areas see **Primary target area**; **Secondary target area**.)

Task.—*a.* Assignment; and operation or mission.

b. A fire mission in gunnery or rifle fire.

c. *British*: The amount of work to be executed by a man, party, or unit during a relief.

Task force.—A temporary tactical unit, composed of elements of one or more arms and services, formed for the execution of a specific mission or operation. *British equivalent:* **Special force.**

Tattoo.—A bugle call sounded at night as a signal that certain lights will be put out; lights out.

Team.—A small group of men working together to operate a gun, radio station, or the like.

Technical sergeant.—A noncommissioned officer of the second grade in the Army, who ranks next above a staff sergeant and next below a master sergeant or first sergeant.

Telegram, daily.—See **Daily telegram.**

Tentacle.—*British term*: A wireless (radio) detachment from an air support signal unit used to request Army air support, and to pass information of interest to both Army and RAF.

Terrain.—An area of ground considered as to its extent and natural feature in relation to its use for a particular operation.

Terrain compartment.—See **Compartment of terrain.**

Territorial Army.—*British term*: A reserve organization of the British Army, corresponding to the National Guard in the United States. At the declaration of war on September 3, 1939, the British Army consisted of the Regular Army, the Territorial Army), and several reserve forces. Soon thereafter all elements were consolidated into a single "British Army" and, except for certain legal differences, the distinctions between these several elements were eliminated.

Theater of operations.—The area of the theater of war in which operations are or may be conducted. It is divided normally into a combat zone and a communications zone. *British equivalent:* **Theatre of operations.**

Theater of war.—The entire area of land, sea, and air which is or may become directly involved in the operations of war. *British equivalent:* **Theatre of war.**

Thompson machine carbine.—*British term:* see **Thompson submachine gun.**

Thompson submachine gun.—A .45 caliber, air-cooled automatic weapon that can be carried and operated by one man. It was used by both the U.S. Army and the British Army. It was familiarly known as the "Tommy gun." *British:* **Thompson machine carbine.**

Third line transport.—*British term:* see **Field Transport, R.A.S.C.** and **Transport (British).**

Time distance.—The distance to a point measured in time. It is found by dividing the ground distance to the point by the rate of march. *British equivalent:* **Time and distance.**

Time fire.—Fire of artillery projectiles which are equipped with time fuzes so that they will burst in the air.

Time fuze.—A fuze that contains a graduated time element in the form of an explosive train or a mechanism similar to a watch, with which the time at which the fuze will function after fired is regulated.

Time interval.—The interval of time between march units or serials, measured from the tail of the one in front to the head of the one in rear. *British equivalent:* **Time allowance.**

Time length.—The time required for a column to pass a given point. *British equivalent:* **Time past a point.**

Time of attack (or "H" hour).—The hour at which the attack is to be launched. The hour designated for the forward movement from the line of departure to begin. *British equivalent:* **Zero hour.**

Time shell.—A shell equipped with a time fuze so that it will burst in the air at a given instant.

Titanium tetrachloride.—A chemical agent used to produce a screening smoke.

TNT.—A high explosive used widely in projectiles and by engineers; trinitrotoluene; trinitrotoluol.

Toxic.—Poisonous. *British equivalent:* Same.

Toxic smoke.—*British term:* see **Irritant smoke.**

Trace.—*a.* A location on a map or on the ground for a trench or other fortification.

    b. A line one a map representing the flight line of an airplane.

c. A path of a tracer bullet. *d.* A path of an element ahead. To follow a marching unit *in trace* is to march behind it over exactly the same route.

d. British: The outline of a work in plan.

Tracer ammunition.—A type of ammunition containing a chemical composition that burns in flight. Tracer ammunition is used for observation of fire, for incendiary purposes, and for signaling. Also called **tracer**.

Tracking.—Observing and plotting the successive positions of a moving target.

Track-laying vehicle.—A vehicle which travels upon two endless tracks, one on each side of the machine. A track-laying vehicle has high mobility and can maneuver, is usually armed and frequently armored, and is intended for tactical use. **Tanks** are one type of track-laying vehicle.

Tractor.—*British usage*: The British Army made a distinction between trucks and lorries, "truck" being used for a load-carrying vehicle of 1 long ton of less, and "lorry" for a load-carrying vehicle of 30-cwt (1.5 long tons) or more. In addition, the term "van" was used for a truck with a fixed top, and "tractor" for a lorry employed to pull or tow anything. Thus all artillery prime movers are designated as tractors.

Tractor-drawn artillery.—Mobile artillery that is moved by a tractor.

Traffic control point.—*British term*: see **Control point**.

Traffic control posts.—*British term*: see **Critical points**.

Traffic map.—*British term*: see **Circulation map**.

Trail.—*a.* The rear part of a gun carriage which connects the piece with the limber or tractor. When the gun is unlimbered, the trail rests on the ground and stabilizes the piece in firing position. *b.* To attach the trail of a gun to the limber.

Trail bridge.—See **Trail ferry**.

Trail ferry.—A raft attached to a cable running across a stream; trail bridge. A trail ferry is attached to the cable so that it can be set at a slant and use the force of the current for power.

Train.—That portion of a unit's transportation, including personnel, operating under the immediate orders of the unit commander primarily for supply, evacuation, and maintenance. It is designated by the name of the unit; such as "1st Infantry Train." Trains are subdivided into "Echelon A" and "Echelon B." Echelon A includes those vehicles of the unit train that are essential for combat, i.e., light maintenance, fuel, lubricants, and ammunition. Echelon B consists of the remaining vehicles of the unit train, i.e., kitchen, baggage, and heavy maintenance. *British equivalent*: **Unit transport, "B" echelon**.

Transfer of fire.—a. The shifting of fire from one target to another, applying the corrections for the first target to the data for the second target. In this meaning, sometimes called **transfer**. b. To fire on a target, based on adjustment on a point whose location in respect to the target is know.

Transfer point.—The point at which control over railway trains, motor convoys, or reinforcements passes from 'one commander to another. *British equivalent*: None.

Transport (U.S.)—A government ship used for carrying troops, supplies, or equipment.

Transport (British).—*a.* The driver, animal, vehicle, and such equipment (e.g. harness) as may be necessary to render the vehicle mobile. (See also **Field Transport, R.A.S.C.**)

   b. That portion of an organized unit, formation, or service of which the primary duty is the transportation of personnel, animals, or material. In the early war period, transport of formations and units was divided into three categories:

   (i) **First line.**—Unit transport;
   (ii) **Second line.**—Transport normally working between **refilling points** and **delivery points**;
   (iii) **Third line.**—Transport normally working between railhead and **refilling points**.

In the late war period, transport was divided into four categories:

   (i) **First line.**—Transport borne on the war establishment of a unit, intended to carry loads for its own use;
   (ii) **Second line.**—Divisional, corps, or army troops companies R.A.S.C., whose duty it is to deliver supplies, etc., to first line transport. Second line transport draws from a maintenance area or railhead. The second line army troops transport may draw from a forward maintenance area, or from a railhead maintenance area;
   (iii) **Army transport companies.**—R.A.S.C. companies under the control of army headquarters, primarily used to carry the requirements of a corps from a Line of Communications terminal to within range of second line transport;
   (iv) **General transport companies.**—R.A.S.C. companies

provided by GHQ troops, primarily on the Line of Communications and in bases.

Transport (or "T") group.—*British term*: see **Group.**

Transportation.—*a.* The movement of personnel and equipment. *British equivalent*: Same.

b. All vehicles or carriers of a unit used for carrying personnel, supplies, and matériel. *British equivalent*: **Transport.**

Transport, supply, and quartering of troops.—*British term*: see **Logistics.**

Traverse.—*a.* To turn a gun to right or left on its mount in pointing.

b. Movement to right of left on a pivot or mount, usually of a gun, but sometimes a piece of machinery.

c. The possible movement of a gun on its mount to right or left, measured as an angle.

d. A projecting mount, or mask of earth or concrete, protecting a trench, emplacement, or other fortification from enfilade fire.

Traversing and searching fire.—Gunfire intended to cover thoroughly a whole area. In traversing and searching fire the gun is gradually traversed in one direction, then the elevation is increased and the gun is traversed in the opposite direction, and so on.

Traversing fire.—Fire distributed in width by moving the gun gradually to the right or left. Traversing fire in the opposite of **searching fire**, which is distributed in depth by successive changes in the elevation of the gun.

Treadway bridge.—A bridge whose roadway is formed by two tracks or treadways.

Trench mortar.—A smooth-bore, muzzle-loading mortar that fired a 3-inch shell; Stoke mortar. The trench mortar was obsolete by World War II and was replaced by the more effective and accurate 81-mm mortar.

Trench warfare.—Warfare in which both sides occupy trenches.

Trial fire.—Deliberate gunfire laid on a fixed point or target to determine the corrections, for firing data. Trial fire is used to prepare for **fire for effect.**

Trial shoot.—*British term*: see **Fire for adjustment** and **Registration.**

Triangle division.—An infantry division made up of three infantry regiments, together with supporting artillery units. It is called a triangular division because it has three infantry regiments instead of four, as in a *square division*.

Trigger guard.—A curved piece of metal on a gun within which the trigger is located and which protects the trigger. Also called a ***guard***.

Trimonite.—A high explosive used as a substitute for TNT as a bursting charge. Trimonite is a mixture of picric acid and monontironaphthalene.

Trinitrotoluene.—The chemical name for TNT.

Trip wire.—*a.* A wire stretched near the ground to trip foot soldiers.

    b. A low outer wire of a double-apron fence or low wire entanglement.

    c. A wire attached to an antitank or antipersonnel mine. Movement of the may may cause detonation of the mine.

Troop.—*a.* An administrative and tactical unit of the Army, in cavalry; subdivision of a squadron of cavalry. A troop corresponds to a company in other branches of the Army, to a battery in artillery, or to a squadron in aviation. *British equivalent:* ***Squadron (cavalry)***.

    b. British: An administrative and tactical unit of the cavalry corresponding to a platoon in the infantry or a section in the artillery.

Trooper.—*a.* A cavalryman; cavalry soldier.

    b. British: A soldier in the lowest grade in the cavalry arm of the Army.

Troop Leading.—The art of leading troops in maneuver and battle. *British equivalent:* Same.

Troop movement.—Transportation or movement of troops from one place to another by any means.

Troop movement by air.—A movement in which troops are moved by means of air transport. *British equivalent:* Same.

Troop movement by marching.—A movement in which foot troops move as such and other units move by their organic transport. *British equivalent:* ***Movement by march route***.

Troop movement by motor.—A movement in which foot troops and all other elements move simultaneously by motor vehicles. *British equivalent:* ***Embusssed movement***.

Troop movement by shuttling.—A movement by motor in which all or a portion of the trucks make successive trips in moving both cargoes and troops. *British equivalent:* Same.

Troop unit.—Any organization of soldiers.

Truce.—An agreement by both sides to stop fighting temporarily, for some special reason, such as to discuss terms of surrender. A truce is generally local. A truce between whole nations is called an ***armistice.***

Truck.—*British usage*: The British Army made a distinction between trucks and lorries, "truck" being used for a load-carrying vehicle of 1 long ton or less, and "lorry" for a load-carrying vehicle of 30-cwt (1.5 long tons) or more. In addition, the term "van" was used for a truck with a fixed top, and "tractor" for a lorry employed to pull or tow anything. Thus all artillery prime movers are designated as tractors.

Truckhead.—See ***Railhead.***

Trunnion.—One of the two pivots supporting a piece of artillery on its carriage and forming the horizontal axis about which the barrel rotates when it is elevated.

Tube.—The main part of a gun, the cylindrical piece of metal surrounding the bore. Tube is frequently used in referring to artillery weapons, and ***barrel*** is more frequently used in referring to small arms.

Turning movement.—A wide enveloping maneuver, passing around the side of the enemy's main force and attacking him from the rear. *British equivalent:* Same.

Turret.—A dome-shaped or cylindrical armored structure containing one or more guns, located on forts, warships, airplanes, and tanks. Most turrets are build so that they can rotate.

Turret defilade.—A condition in which a tank, except for its turret, is hidden from the enemy by an intervening hill or other mask. See also ***Hull defilade.***

Turret down.—*British term*: A condition in which the tank commander, from a position in or on the tank, can observe and correct fire, but when the tank itself is concealed from the potential target.

Ultimatum.—A final demand that another nation or military force agree to meet certain conditions, carrying with it a threat of military action if the demand is not met.

Unauthorized belligerents.—Men or units not recognized as lawful combatants under the rules of land warfare, and so not entitled to be treated as prisoners of war if captured.

Unconditional surrender.—An act of yielding completely, without reservation or terms.

Uncontrolled mosaic.—An assembly of two or more overlapping vertical photographs accomplished by the matching of photographic detail only, without the benefit of a framework of control points. *British equivalent*: Same.

Uncover.—To expose or leave unprotected by movement or maneuver.

Unit.—*a.* A military force having a prescribed organization.

b. British: An organization of a single arm or service operating both tactically and administratively under a single commander, whereas a ***formation*** being a combination of units of different arms and services to the strength of a brigade or more.

Unit distribution.—The delivery of Class I supplies to the regimental (or similar unit) kitchen areas by the quartermaster.

United States Army.—The permanent military force of the United States. The term is generally used to refer to the Regular Army, although in includes the other normal, peacetime components, such as the National Guard of the United States and the Organized Reserves. It differs from the ***Army of the United States,*** which is a temporary military organization in time of war, including the permanent components as well as the temporary components such as Selective Service personnel, and the National Guard while it is in federal service.

Unit in support.—*British term*: see ***Supporting unit.***

Unit load.—A term used to indicate method of loading vehicles; supplies required for a particular unit being loaded as required on one or more vehicles. *British equivalent*: None.

Unit loading.—Method of loading which gives primary consideration to the availability of the troops for combat purposes on detraining or landing, rather than utilization of railroad or ship space.

    Combat unit loading.—Method of loading in which certain units are completely loaded on one train or ship with at least their essential combat equipment and supplies immediately available for detraining or debarkation with the troops, together with the animals and motors for the organization when this is practicable.

    Convoy unit loading.—Method of loading in which the troops with their equipment and supplies are loaded on transports of the same convoy, but not necessarily on the same transport.

    Organizational unit loading.—Method of loading in which organizations, with their equipment and supplies, are loaded on the same train or transport, but not so loaded as to allow detraining or debarkation of troops and their equipment simultaneously.

Unit mile of gasoline.—The amount of gasoline in gallons required to move every vehicle of the unit one mile. *British equivalent*: None.

Unit of fire.—The quantity in rounds or tons of ammunition, bombs, grenades, and pyrotechnics which a designated organization or weapon may be expected to expend on the average in one day of combat. *British equivalent*: None.

Unit replacement.—The system of repair by which an unserviceable unit assembly is replaced by a like, serviceable unit assembly. *British equivalent*: **Replacement of assemblies**.

Unit of supply.—A measure for ammunition supply within a theater, based on experience in the theater. It represents a specific number of rounds per weapon, which varies with the types and calibers of the weapons. The unit of supply is not synonymous with the term ***day of supply***.

Unit reserves.—The prescribed quantities of supplies carried as a reserve by a unit.

Unit transport, "B" echelon.—*British term*: see ***Train***.

Unit "under command."—*British term*: see ***Attached unit***.

Universal carrier.—*British term*: A type of British armored carrier. A Universal carrier fitted with rests for a Bren gun and a Boys anti-tank rifle is commonly known as a Bren gun carrier.

Unmask.—To make a tactical move so as to get out of the way and leave a clear field of fire open for friendly troops.

Urgent call.—A telephone call believed by the calling party to be more important than any call which might be in progress. *British equivalent*: **Priority call**.

Urgent message.—A message requiring the greatest speed in handling. *British equivalent*: **Immediate (or most immediate) message**.

Usable rate of fire.—The normal rate of fire of a gun in actual use, measure in units of shots per minute. The usable rate of fire is considerably less than a gun's maximum rate of fire, which is a theoretical value based on the purely mechanical operation of a weapon.

Van.—*British usage*: The British Army made a distinction between trucks and lorries, "truck" being used for a load-carrying vehicle of 1 long ton or less, and "lorry" for a load-carrying vehicle of 30-cwt (1.5 long tons) or more. In addition, the term "van" was used for a truck with a fixed top, and "tractor" for a lorry employed to pull or tow anything. Thus all artillery prime movers are designated as tractors.

Van guard.—*British term*: see **Advance party**.

Vehicle park.—An area used for the purpose of parking vehicles. The place where vehicles of a unit are parked.

Verbal order.—*British term*: see **Oral order**.

Vertical aerial photograph.—An aerial photograph made with a camera whose optical axis is at or near the vertical.

Vertical envelopment.—An envelopment of the enemy from the air; a tactical maneuver in which aircraft attack the rear and flanks of a force, in effect cutting it off or encircling it.

Vertical interval.—See **Contour interval**.

Very light.—A colored signal flare fired from a special pistol; Very signal light.

Very pistol.—A special pistol used to fire colored signal flares.

Vesicant.—A chemical agent which is readily absorbed or dissolved in both the exterior and interior parts of the human body, followed by the production of inflammation, burns, and destruction of tissue. *British equivalent:* Same.

Vickers medium (heavy) machine gun.—A British .303 caliber machine gun that was the basic weapon of the (heavy) machine-gun battalion. Although a "heavy" machine gun by U.S. terminology, it was classified as a "medium" machine gun by the British. Water-cooled, recoil-operated, belt0-fed. The mounting consisted of a crosshead elevating gear and a socket mounted on three legs.

Vision slit.—Any narrow opening or slit in armor through which to look, especially one in a tank or other armored vehicle.

Visual signals.—Signals conveyed through the eye; they include signals transmitted by flags, lamps, panels, heliograph, pyrotechnics, hand, and arm. *British equivalent:* Same.

Vital point.—*British term*: see **Key point**.

V-mail.—Mail to or from members of the armed forces serving overseas, written on a special blank. V-mail is almost always photographed on microfilm for transportation, in order to reduce its weight and size, and is enlarged and printed before delivery.

"Walkie talkie."—(*colloquial*). A radio set which may be carried and operated for both receiving and sending by one man. *British equivalent:* Same.

WAC.—Women's Army Corps. The members of the Women's Army Corps hold the same comparative ranks as soldiers of the Army of the United States.

Walking wounded.—A sick or wounded person who can walk from the place where he became a casualty to the place where he can receive medical treatment; ambulant case.

War Department intelligence.—The military intelligence produced under the direction of the War Department General Staff in peace and in war. *British equivalent*: **War Office intelligence**.

War Establishments.—*British term*: see **Tables of Organization (T/O)**.

War diary.—A log or daily account of events, kept by an organization in a campaign. See also **Journal**.

War material.—All tangible requirements for war other than personnel and animals.

Warning net.—Any system of observation posts linked together by a communication system to give mutual warning of enemy movement; in particular, a group of radio sets tuned to the same frequency so that all units will hear a warning sent out by any one.

Warning order.—An order issued as a preliminary to another order, especially for a movement, which is to follow; it may be a message or a field order, and may be either written, dictated, or oral. The purpose is to give advance information so that the commanders may make necessary arrangements to facilitate the execution of the subsequent field order. *British equivalent*: Same.

War of masses.—Warfare in which the number of men, rather than the power of mechanized equipment and mobile weapons, is decisive.

War of movement.—Mobile warfare in which the opposing sides seek to seize and hold the initiative by the use of maneuver, organization of fir, and utilization of terrain, as contrasted with **war of position** or **position warfare**, in which the defensive is confined to a fixed position. Also called **mobile warfare**.

War of position.—Warfare in which the defensive is confined mainly to fixed positions. The defense is aimed chiefly at keeping the enemy out of strategic areas and at forcing him to exhaust his combat power in assaults against well-established positions. Also called **position warfare**.

War Plans Division.—A subdivision of the War Department General

Staff which makes plans to employ the military forces in various possible situations, in peacetime. In time of war, the War Plans Division is called **Operations Division**, and is in charge of the working out of strategical operations.

Warrant officer.—*a.* An officer in the U.S. Army or U.S. Navy who is neither a commissioned officer nor an enlisted man, holding his grade by authority of appointment or warrant. A warrant officer ranks next above a cadet. or midshipman or noncommissioned officer, and next below a second lieutenant or ensign. In the U.S. Army there were two grades: chief warrant officer and warrant officer, junior grade. In the U.S. Navy there were also two grades: commissioned warrant officer and warrant officer.

    b. British: A member of the highest-ranking group of noncommissioned ranks, holding a King's Warrant. In the British Army there were two grades: Regimental sergeant-major, who is ranked as warrant officer, class I; and regimental quarter-master sergeant and company sergeant-major, who are ranked as warrant officer, class II. The grade of warrant officer, class III was discontinued soon after the outbreak of World War II.

War strength.—The minimum strength in personnel and equipment with which a unit can operate effectively under prolonged war conditions.

Wave.—One of a series of lines of foragers, mechanized vehicles, skirmishers, or small columns into which an attack unit is deployed in depth.

Weapon pit.—*British term*: see **Fox hole.**

White phosphorus.—A yellow, waxy solid that burns easily when not kept under water. White phosphorus is used as an incendiary and smoke-producing agent.

Wide envelopment.—An enveloping maneuver that starts from the enemy position and usually is directed at an objective far in the rear of the enemy front lines.

"Wilco."—A word used to show that an operator has received a radio telephone order and will carry it out. It stands for "will comply" or "will carry out orders.

Wing.—*a.* An administrative and tactical unit in the U.S. Army Air Forces. A wing is larger than a group, and smaller than a command. Wings usually contain only one type of aviation, although composite wings may be formed where the situation demands. It is usually commanded by a brigadier general. *British equivalent*: **Group**

(which is composed of two or more wings).

    b. British: The smallest formation of the R.A.F. Whenever practicable, each wing is composed of squadrons carrying out similar duties and designated accordingly, e.g. bomber wing, fighter wing. (*Note that the organizational hierarchy of R.A.F. is Group, Wing, Squadron, compared to the U.S.A.A.F hierarchy of Wing, Group, Squadron.*)

    c. A flank unit; that part of a military force to the right or left of the main body.

Wire entanglement.—An obstacle of barbed wire, erected in place on pickets, or constructed in rear of the site and brought up and placed in position. *British equivalent*: Same.

Wireless-telegraphy.—*British term*: see **Radiotelegraphy (or radio (key))**.

Wire telegraphy.—Telegraph communication by code over a wire circuit.

Wire telephony.—Telephone communication by voice over a wire circuit.

Withdrawal.—*a.* An operation of breaking off combat; retirement from action; planned, orderly movement to the rear, as contrasted to a retreat or rout.

    b. British usage: encompasses **Retreat**, **Retirement**, and **Withdrawal from action.**

Withdrawal from action.—The operation of breaking off combat with a hostile force.

Women's Army Auxiliary Corps (WAAC).—An organization of women for noncombatant service with the Army. It was replaced by the **Women's Army Corps** in September, 1943.

Women's Army Corps (WAC).—An organization for noncombatant service with the Army. The members of the Women's Army Corps hold the same comparative ranks as soldiers of the Army of the United States. It replaced the **Women's Army Auxiliary Corps** in September, 1943.

Writer.—The originator of a message.

X casualty.—*British term*: A vehicle breakdown due to a temporary stoppage only, and repairable by the crew of the vehicle without other assistance.

X List.—*British term*: A list maintained for each corps on which all personnel of a corps, in a theater of operations, are accounted for which are not serving with units of that corps.

X-X line.—An imaginary line inside the enemy position parallel to the front line, which marks the limit in depth of the responsibility of division artillery. Targets farther back behind enemy lines are commonly the responsibility of corps artillery and army artillery.

Y casualty.—*British term*: A vehicle breakdown that requires repair personnel, and likely to be repairable but unit fitters and light aid detachments of second-line workshops.

Y-Y line.—An imaginary line inside the enemy position parallel to the front line, which marks the division between zones of responsibility for subdivisions of corps artillery. The Y-Y line lies between the ***X-X line*** and the ***Z-Z line***.

Z casualty.—*British term*: A vehicle break down requiring extensive repair or replacement, involving evacuation.

Zero hour.—The hour set for an attack or other operation to begin. See also ***Time of attack (or "H" hour)***.

Zero-in.—To adjust the site settings of a rifle by calibrating firing on a standard range with no wind blowing.

Zone.—*a.* Any tactical area of importance, generally parallel to the front, such as a fortified area, a defensive position, a combat zone, a traffic control zone, etc.

    b. A strip of several bands or belts of wire entanglements placed in depth.

    c. An area in which projectiles will fall for a given propelling charge, when the elevation is varied between the minimum and maximum.

Zone defense (or defense in depth).—*A* form of defense which includes several successive battle positions, more or less completely organized. Zone defense differs from ***position defense***, in which the defense of the whole area is conducted from one main center of resistance. *British equivalent*: ***Defence in depth***.

Zone fire.—Fire that completely covers the region in which the target is situated. In zone fire, different guns of a battery use different ranges, or elevations, so that they sweep an entire zone. *British equivalent*: Same.

Zone of action.—A zone designated by boundaries in an advance or a retrograde movement within which the unit operates and for which it is responsible. *British equivalent*: ***Front*** (i.e., company, battalion, etc.).

Zone of advance.—A designated geographical region through which a military unit is to advance. The zone of advance is an area of responsibility marked off by boundaries on each side. If possible, the boundaries of a zone of advance are easily identified ground features, such as streams or ridges.

Zone of fire.—The area within which a unit is to be prepared to deliver fire.

Zone of the interior.—The area of the national territory not included in the theaters of operations. The zone of the interior is organized to furnish manpower and munitions to the armed forces. *British equivalent*: None.

Z-Z line.—An imaginary line farther inside the enemy position than the **X-X line** or **Y-Y line**, and parallel to the front line, which marks the limit in depth of the responsibility of corps artillery. Targets farther behind enemy lines than the Z-Z line are usually the responsibility of army artillery.

DIAGRAMS.

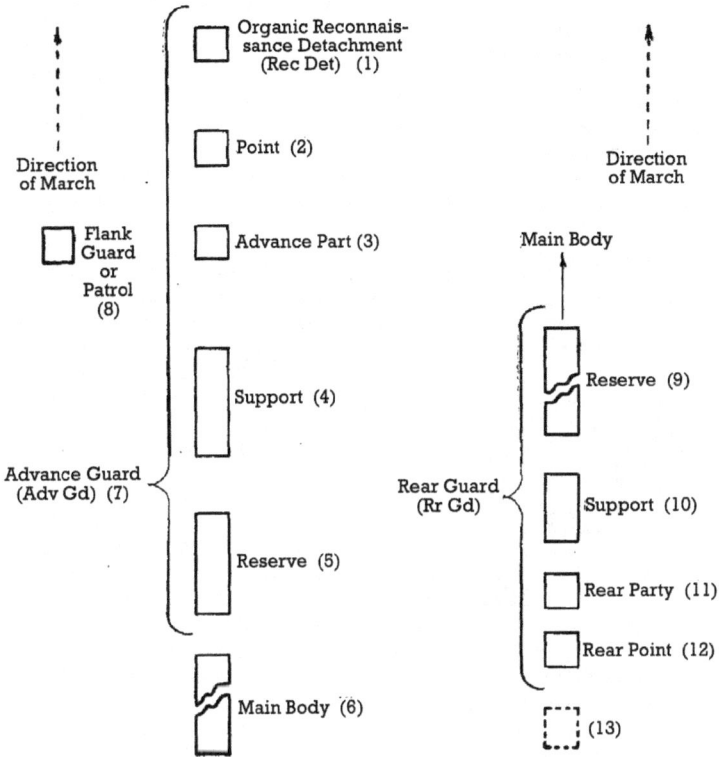

Figure 1.— March dispositions.
(U.S. and equivalent British terminology).

British equivalents.

Route column:
- (1) Mobile troops (reconnaissance (recce) unit).
- (2), (3), (4) Van guard.
- (5) Main guard.
- (6), (8) (Same.)
- (7) Advanced guard.

Rear guard:
- (9), (10) Main body of rear guard.
- (11), (12) Rear parties of rear guard.
- (13) Rear guard mobile troops.

① U.S. ROUTE COLUMN   ② U.S. REAR GUARD

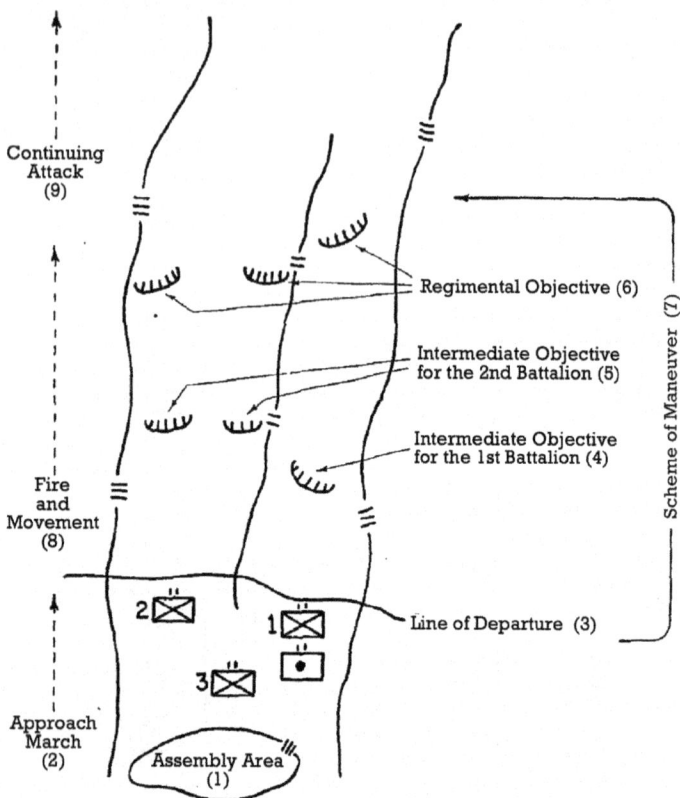

British equivalents.

(1) Assembly position.	(4) (Same.)	(7) Plan of attack.
(2) (Same.)	(5) (Same.)	(8) (Same.)
(3) Start(ing) line (SL).	(6) (Same.)	(9) Exploitation.

Figure 2.— Regiment in attack.
(U.S. and equivalent British terminology).

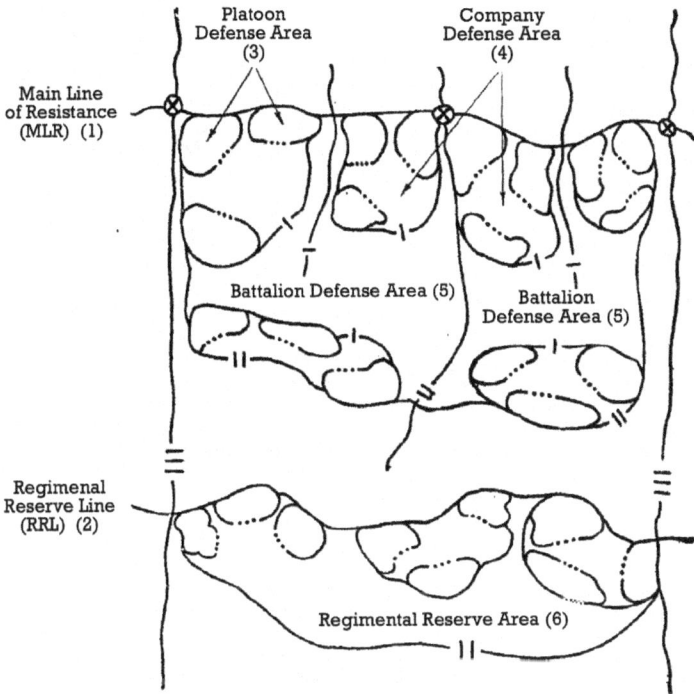

British equivalents.

(1) (Line of) forward defended locali-
    ties (FDL's).
(2) Brigade reserve position.
(3) Platoon sector.

(4) Company sector.
(5) Battalion sector.
(6) Brigade reserve area.

Figure 3.— Regiment in defense.
(U.S. and equivalent British terminology).

FRONT LINE

1st DIV. | 2nd DIV. | 3rd DIV. | 4th DIV. | 5th DIV. | 6th DIV. | 7th DIV. | 8th DIV. | 9th DIV.

COMBAT ZONE

FIRST CORPS

SECOND CORPS

THIRD CORPS

ARMY SERVICE AREA

COMMUNICATIONS ZONE

ADVANCE SECTION

INTERMEDIATE SECTION

BASE SECTION

BOUNDARY LEGEND:

—— X X —— DIVISIONS
—— X X X —— CORPS
—— X X X X —— ARMY
—— O O O —— SECTION, COMMUNICATIONS ZONE
—— O O O O —— COMMUNICATIONS ZONE

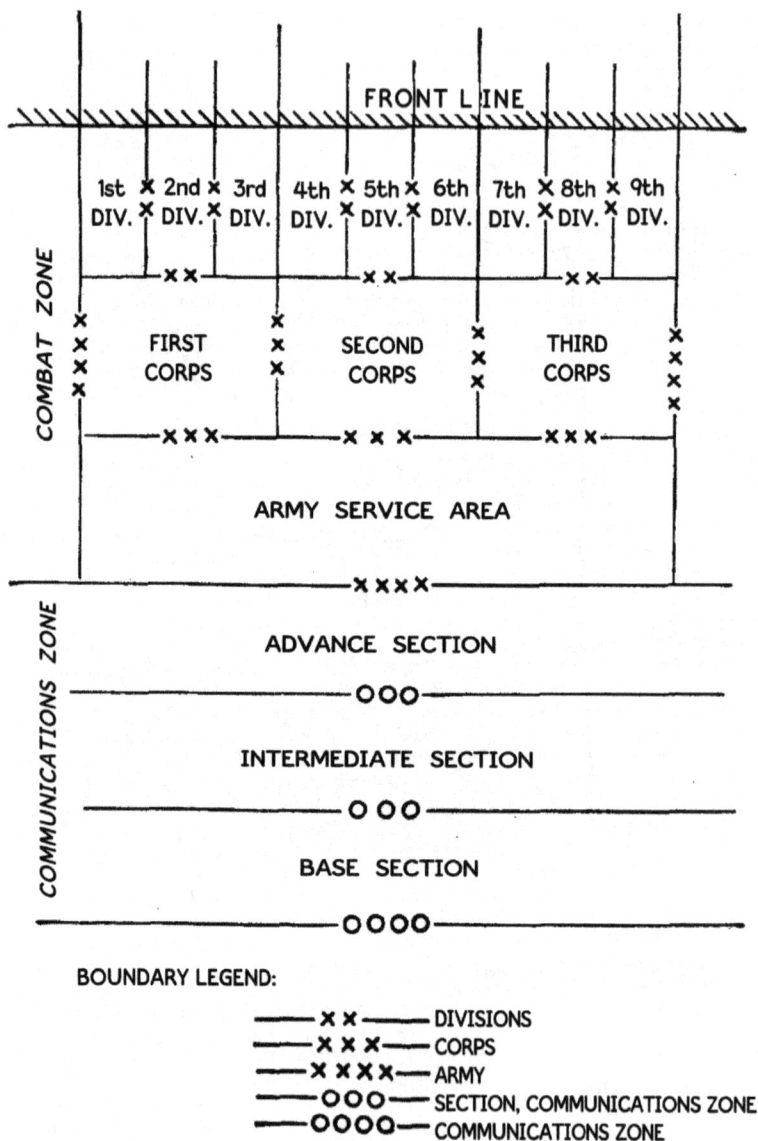

Figure 4.— Organization of a U.S. theater of operations.

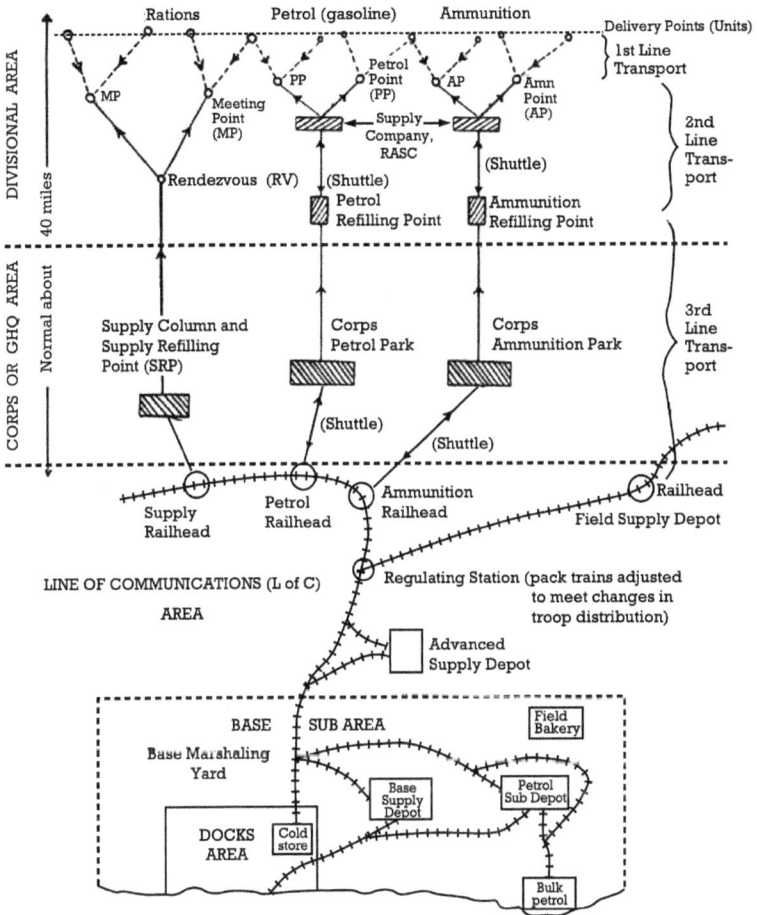

Figure 5.— British supply system.

DESIGNATIONS OF COMBAT UNITS

Table 1.— Designations of U.S. units.

Arm	Units			
	Colonel's Command	Lt. Colonel's or Major's Command	Captain's Command	Lieutenant's Command
Infantry	Regiment	Battalion	Company	Platoon
Armor	Regiment	Battalion	Company	Platoon
Cavalry	Regiment	Squadron	Troop	Platoon
Artillery	Regiment	Battalion	Battery	Platoon
Engineers	Regiment	Battalion	Company	Platoon
Signals	—	Battalion	Company	Platoon
Army Air Forces				
Brigadier General's Command	Colonel's Command	Lt. Colonel's or Major's Command	Captain's Command	Lieutenant's Command
Wing	Group	Squadron	Flight	1 airplane

Table 2.— Designations of British units.

Arm	Units			
	Brigadier's Command	Lieut. Col.'s Command	Major's or Capt.'s Command	Subaltern's Command
Infantry	Brigade	Battalion	Company	Platoon
Tanks	—	Battalion	Company	Section
Cavalry	Brigade	Regiment	Squadron	Troop
		Cavalry Armoured Car Regiment	Squadron	Section
Artillery	varies	Brigade	Battery	Section
Engineers	—	Divisional Engineers	Field Company	Section
			Squadron	Troop
		Searchlight Battalion	Company	Section
Signals	—	varies	Company Cavalry Divisional Signals	Section Troop

Royal Air Force

Group Captain's Command	Wing Commander's Command	Squadron Leader's Command	Flight Lieutanat's Command	Flying Officer's Command
Group	Wing	Squadron	Flight	1 aeroplane

COMBAT UNITS

Table 3.— Designations of aviation tactical units.

	Large Unit	Intermediate Unit	Basic Unit[1]
U.S. Army Air Forces	Wing	Group	Squadron[2]
U.S. Navy		Carrier Air Group[3] *or* Patrol Wing[4] [5]	Squadron[2]
Royal Air Force	Group	Wing	Squadron[2]
Royal Navy		Carrier Air Group	Squadron[2]
French Air Force	Escadre *or* Groupement[6]	Groupe[2]	Escadrille
French Navy		Flotille	Escadrille[2]
Germany	Geschwader	Gruppe[2]	Staffel
Italy	Stormo	Gruppo[2]	Squadriglia
Japanese Army Air Service	Hikōdan *(Air Brigade)*	Hikō Sentai[2] *(Air Regiment)*	Hikō Chūtai *(Air Company)*
Japanese Navy		Kōkū Sentai *(Air Regiment)[4]*	Kōkū-tai *(Air Squad)[3]* or Buntai *(Squad)[4]*
Soviet Union	Aviapolk *(Aviation Regiment)*	Eskadrilya *(Squadron)*	Aviaotryad *(Aviation Squad)*

Notes.—Direct equivalencies are difficult as the number of aircraft in these tactical units varied considerably between countries, between types of aircraft (fighters, bombers, reconnaissance,, etc.), and between services (land-based, naval). There were also differnces between countries in the number of subordinate units under each of the more senior units.

[1] Smallest operational unit.

[2] Basic combat unit for both administrative and operational purposes.

[3] Carrier-based squadrons only.

[4] Sea-plane and land-based squadrons only.

[5] On 1 November 1942, patrol wings were redesignated fleet air wings to permit patrol aviation to be utilized within the task force principle, to include a variety of commands and a variable number of squadrons necessary to accomplish a particular objective or mission.

[6] The *Escadre* was a pre-war organization comprising of two or three *Groupes* of the same type (fighter, bomber, reconnaissance, etc.). At the outbreak of the war, the *Escadres* were broken up, with the *Groupes* being reorganized into *Groupements* of various sizes, however, with the *Groupes* retaining their old *Escadre* designations. For example, GC III/2 (3rd *Groupe de Chasse* (fighter group) of the 2nd *Escadre*) retained this designation when it was reassigned to *Groupement 25*.

ARMS AND SERVICES

Table 4.— U.S. and British arms and services.

Branches of the U.S. Army	British Army *Arms of the Service*[1] (in order of precedence)
Arms[2]	*The Arms*
Infantry	Cavalry
Cavalry	Royal Armoured Corps[4]
Field Artillery Corps	Royal Regiment of Artillery[5]
Coast Artillery Corps	Corps of Royal Engineers
Air Corps	Royal Corps of Signals
Corps of Engineers[3]	Infantry
Signal Corps[3]	Reconnaissance Corps
Services	*The Services*
Adjutant General's Department	Royal Army Chaplains's Department
Chemical Warfare Service[6]	Royal Army Service Corps
Corps of Chaplains	Royal Army Medical Corps
Finance Department	Royal Army Ordnance Corps
Inspector General's Department	Royal Electrical and Mechanical
Judge Advocate General's Depart- ment	Engineers Royal Army Pay Corps
Medical Department	Royal Army Veterinary Corps
Ordnance Department	Army Educational Corps
Quartermaster Corps	The Army Dental Corps
Transportation Corps	Pioneer Corps
	Intelligence Corps
	Army Catering Corps
	Army Physical Training Corps
	Corps of Military Police
	Military Provost Staff Corps
	Queen Alexandra's Imperial Military Nursing Service
	Auxiliary Territorial Service
	Officers' Training Corps

Notes:

[1] All branches of the British Army, taken collectively, are called the "Arms of the Service."

[2] The Armored Force, established in 1940 and renamed the Armored Command in 1943, did not have the status as a separate branch of the U.S. Army.

[3] Also has service functions.

[4] All except two of the cavalry regiments were mechanized and constituted a part of the Royal Armoured Corps..

[5] The Royal Regiment of Artillery retains the name "regiment" for traditional reasons.

[6] Also has combat units.

RELATIVE MILITARY RANKS

Table 5.— U.S. military ranks.

Army	Marine Corps	Navy
General of the Army	(no equivalent)	Fleet Admiral
General	(no equivalent)	Admiral
Lieutenant General	Lieutenant General	Vice-Admiral
Major General	Major General	Rear Admiral
Brigadier General	Brigadier General	Commodore
Colonel	Colonel	Captain
Lieutenant Colonel	Lieutenant Colonel	Commander
Major	Major	Lieutenant Commander
Captain	Captain	Lieutenant
1st Lieutenant	1st Lieutenant	Lieutenant (Junior Grade)
2nd Lieutenant	2nd Lieutenant	Ensign
Chief Warrant Officer	Commissioned Warrant Officer	Chief Warrant Officer
Warrant Officer (Junior Grade)	Warrant Officer	Warrant Officer
Flight Officer (Army Air Forces)		
Cadet or Aviation Cadet	(no equivalent)	Midshipman
Master Sergeant *or* 1st Sergeant[1]	Sergeant Major or Master Technical Sergeant	Chief Petty Officer
Technical Sergeant	1st Sergeant or Quartermaster Sergeant / Gunnery Sergeant or Technical Sergeant	Petty Officer 1st Class
Staff Sergeant or Technician 3rd Grade[2]	Platoon Sergeant or Staff Sergeant	Petty Officer 2nd Class
Sergeant or Technician 4th Grade	Sergeant	Petty Officer 3rd Class
Corporal or Technician 5th Grade	Corporal	Seaman 1st Class
Private, First Class	Private, First Class	Seaman 2nd Class
Private	Private	Apprentice Seaman

Notes:
   [1] Prior to September 22, 1942, 1st Sergeants were ranked with Technical Sergeants.
   [2] Technicians were generally not addressed as such, but rather as the equivalent line NCO rank in his grade.

A GLOSSARY OF WORLD WAR II MILITARY TERMS

Table 6.— British military ranks.

Army	Royal Air Force	Royal Navy[1]
Field Marshal	Marshal of the Royal Air Force	Admiral of the Fleet
General	Air Chief Marshal	Admiral
Lieutenant General	Air Marshal	Vice-Admiral
Major General	Air Vice-Marshal	Rear Admiral
Brigadier[2]	Air Commodore	Commodore (1st and 2nd class)
Colonel	Group Captain	Captain
Lieutenant Colonel	Wing Commander	Commander
Major	Squadron Leader	Lieutenant Commander
Captain	Flight Lieutenant	Lieutenant
Lieutenant	Flying Officer	Sub-Lieutenant
Second-Lieutenant....	Pilot Officer / Acting Pilot Officer (but junior to Navy and Army Ranks)	Acting Sub-Lieutenant / Commissioned Warrant Officer
†Conductor, Royal Army Ordnance Corps; †Master Gunner, 1st class; †1st class Staff Serjeant-Major	(no equivalent)	Warrant Officer (but senior to Army ranks) Midshipman (but junior to Army Ranks)
Warrant Officer, Class I, except those marked † above	Warrant officer	(no equivalent)
Warrant Officer, Class II	(no equivalent)	(no equivalent)
Squadron Quarter-Master-Corporal (Household Cavalry); Squadron, Battery, or Company Quarter-Master Serjeant; Colour-Serjeant; Staff-Corporal (Household Cavalry); or Staff-Serjeant	Flight Sergeant	Chief Petty Officer

Table 6.— British military ranks (continued).

Corporal-of-Horse (Household Cavalry) or Serjeant *Lance-Serjeant*[3]	Sergeant	Petty Officer
Corporal or Bombadier *Lance-Corporal or Lance-Bombadier*[3]	Corporal	Leading Seaman
Trooper, gunner, sapper, signalman, driver, guardsman, fusilier, or private	Leading Aircraftman / Aircraftman, 1st class / Aircraftman, 2nd class	Able Seaman / Ordinary Seaman

Notes:

[1] The Royal Marines are part of the Royal Navy but use the Army names of ranks. Warrant officers, Royal Marines, except Sergeant-Majors (whose rank is the equivalent of Warrant Officer, class I, in the Army), rank with Warrant Officers, Royal Navy. Commissioned officers of the Royal Marines rank, according to seniority, with officers of the Army of the same titles. Commissioned officers from Warrant rank, Royal Marines, rank with commissioned officers from Warrant rank, Royal Navy.

[2] "Brigadier" is the proper title. It is temporary for the job, holding the appointment only while commanding a brigade or performing other duties for which the appropriate rank is a brigadier. Although a brigadier corresponds to the rank of brigadier general in the U.S. Army, it is improper to address such and officer as "General." The British rank of brigadier-general was abolished in 1920.

[3] Lance-serjeants and lance-corporals/lance-bombadiers are appointments, not ranks, and are given to corporals and privates, respectively, who have qualified and been recommended for further promotions when vacancies occur. They rank with corporals and privates, respectively, but are senior to those ranks and to the corresponding ranks in the Royal Navy and Royal Air Force. Note also that the British spelling of the rank of sergeant during World War II was "serjeant" in the Army but

A GLOSSARY OF WORLD WAR II MILITARY TERMS

Table 7.— British warrant officer appointments.

Rank	Appointment
Warrant Officer, Class I	Conductor, Royal Army Ordnance Corps Master-gunner, 1st class Staff serjeant-major Master-gunner, 2nd class Garrison serjeant-major Corporal-major, Household Cavalry Regimental serjeant-major
Warrant Officer, Class II	Master-gunner, 3rd class Garrison quartermaster-serjeant Regimental quartermaster-corporal, Household Cavalry Regimental quartermaster-serjeant Quartermaster-serjeant Staff quartermaster-serjeant

Note.—Warrant officers in the British Army were generally referred to by their appointment, which described their specific duties, rather than by their rank. The table above includes the more important appointments and their order of precedence within rank.

Table 8.— French military ranks.

Grade	French Army (*Armée de terre*) and French Air Force (*Armée de l'air*)	French Navy (*Marine nationale*)
—	Maréchal de France[1]	Amiral de France[1]
Field Marshal	—	Amiral de la Flotte[2]
General	Général d'armée[3]	Amiral
Lt. General	Général de corps d'armée[3]	Vice-amiral d'Escadre
Major General	Général de division[3]	Vice-amiral
Brigadier General	Général de brigade[3]	Contre-amiral
Colonel	Colonel	Capitaine de vaisseau
Lt. Colonel	Lieutenant-colonel	Capitaine de frégate
Major	Commandant[4]	Capitaine de corvette
Captain	Capitaine	Lieutenant de vaisseau
1st Lieutenant	Lieutenant	Enseigne de vaisseau de 1re classe
2nd Lieutenant	Sous-lieutenant	Enseigne de vaisseau de 2e classe
Officer Candidate	Aspirant	Aspirant
Student Officer	Éléve-officier	Éléve-officier
Chief Warrant Officer	Adjudant-chef	Maître principal
Warrant Officer	Adjudant	Premier-maître
Sergeant-major	Sergent-chef	Maître
Sergeant	Sergent	Second-maître
Corporal-major	Corporal-chef	Quartier-maître de 1re classe
Corporal	Corporal	Quartier-maître
Private	Soldat	Matelot

Notes:

[1] The titles of Marshal of France (*Maréshal de France*) and Admiral of France (*Amiral de France*) are dignities, not ranks, that are awarded for exceptional service. The last award of Admiral of France was made in 1869. There is no Air Force equivalent.

[2] The rank of Admiral of the Fleet (*Amiral de la Flotte*) was created in 1939 so that Admiral Darlan, the Commander-in-Chief of the French Navy, would not have an inferior rank to that of his counterpart in the British Royal Navy.

[3] The full titles of equivalent French Air Force ranks are: *Général d'armée aérienne*, *Général de corps d'armée aérien*, *Général de division aérienne*, and *Général de brigade aérienne*, respectively.

[4] *Chef de bataillon* in the infantry; *chef d'escadron* in the cavalry and engineers.

Table 9.— German military ranks.

Grade	German Army (Heer)	Waffen-SS
Field Marshal	Generalfeldmarschall	Reichsführer-SS
General	Generaloberst	Oberstgruppenführer
Lieutenant General	General der (arm)[1]	Obergruppenführer
Major General	Generalleutnant	Gruppenführer
Brigadier General	Generalmajor	Brigadeführer
Senior Colonel	—	Oberführer
Colonel	Oberst	Standartenführer
Lieutenant Colonel	Oberstleutnant	Obersturmbannführer
Major	Major	Strmbannführer
Captain	Hauptman or Rittmeister (cavalry)	Hauptsturmführer
1st Lieutenant	Oberleutnant	Obersturmführer
2nd Lieutenant	Leutnant	Untersturmführer
Staff Sergeant	Stabsfeldwebel	Sturmscharfürher
First Sergeant	Hauptfeldwebel	Hauptscharführer
Senior Sergeant	Oberfeldwebel	Oberscharführer
Sergeant	Feldwebel	Scharführer
Junior Sergeant	Unterfeldwebel	Unterscharfürher
Corporal	Unteroffizier	
Staff Lance Corporal	Stabgefreiter	
Senior Lance Corporal	Obergefreiter	SS-Rottenführer
Lance Corporal	Gefreiter	SS.-Sturmmann
Private 1st Class	Obersoldat[2]	SS-Obersoldat[2]
Private	Soldat[2]	SS-Soldat[2]

Notes:

  [1] Artillery: General der Artillerie; Infantry: General der Infanterie; Cavalry: General de Kavallerie; Panzer troops: General der Panzertruppen, Engineers: General der Pioniere, etc.

  [2] "Soldat" is the collective form. It is replaced by a specific form in each arm by:

Grenadier (shütze)	Infantry (rifleman)
Shütze	Tanks, antitank
Reiter......................	Cavalry
Kanonier....................	Artillery
Pionier.....................	Engineers
Funker.....................	Signal
Fahrer	Transport (horse-drawn)
Kraftfahre..................	Motor transport
Sanitätssoldat...............	Medical

In addition, tank, antitank, and armored infantry also have these two ranks prefixed with "Panzer-" to form Panzer-Obershütze, Panzer-Obergrenadier, Panzer-Shütze, and and Panzer-Grenadier.

Table 9.— German military ranks (continued).

Grade	German Air Force (*Luftwaffe*)	German Navy (*Kriegsmarine*)
Field Marshal	Generalfeldmarschall	Grossadmiral
General	Generaloberst	Generaladmiral
Lieutenant General	General der Flieger	Admiral
Major General	Generalleutnant	Vizeadmiral
Brigadier General	Generalmajor	Konteradmiral
Senior Colonel	—	Kommodore
Colonel	Oberst	Kapitän zur See
Lieutenant Colonel	Oberstleutnant	Fregattenkapitän
Major	Major	Korvettenkapitän
Captain	Hauptman	Kapitänleutnant
1st Lieutenant	Oberleutnant	Oberleutnant zur See
2nd Lieutenant	Leutnant	Leutnant zur See
Staff Sergeant	Stabsfeldwebel	Stabsoberfeldwebel
First Sergeant	Hauptfeldwebel	Oberfeldwebel
Senior Sergeant	Oberfeldwebel	Stabsfeldwebel
Sergeant	Feldwebel	Feldwebel
Junior Sergeant	Unterfeldwebel	Obermaat
Corporal	Unteroffizier	Maat
Senior Staff Lance Corporal	—	Oberstabsgefreiter
Staff Lance Corporal	—	Stabsgefreiter
First Lance Corporal	Hauptgefreiter	Hauptgefreiter
Senior Lance Corporal	Obergefreiter	Obergefreiter
Lance Corporal	Gefreiter	Gefreiter
Private	Flieger	Matrose

Table 10.— Italian military ranks.

Grade	Royal Italian Army (*Regio Escercito*)	Fascist Militia (*Milizia Volontária per la Sicureza Nazionale*)
Commander-in-Chief..	{ Maresciallo dell'Impero[1] { Maresciallo d'Italia	— —
General	Generale d'Armata Gen. designate d'Armata	Comandante Generale —
Lieutenant General	Gen. di Corpo d'Armata	Luogatenete Gen. Capo di Stato Maggiore
Major General	Generale di Divisione[2]	Luogatenete Generale
Brigadier General	Generale de Brigata[2]	Console Generale
Colonel	Colonello	Console
Lieutenant Colonel	Tenente Colonnello	Primo Seniore
Major	Maggiore	Seniore
Captain	Capitano[3]	Centurione
1st Lieutenant	Tenente[3]	Capo Manipolo
2nd Lieutenant	Sotto tenente	Sette Capo Manipolo
Officer candidate	Aspirante[4]	Asp. Sette Capo Manipolo
(no U.S. equivalent)	Maresciallo maggiore[5] Maresciallo capo Maresciallo ordinário	Primo Aiutante Aiutante Capo Aiutante
Sergeant-major	Sergente maggiore	Primo Capo Squadra
Sergeant	Sergente	Capo Squadra
Lance-corporal	Caporale[6]	Vice Capo Squadra
Private, first class	Appuntato	Camicio Nero
Private	Soldato	Aviere

Notes:

[1] Held by the King and Mussolini.

[2] The title of lieutenant general (*tenente generale*) and major general (*maggiore generale*), equivalent to U.S. major general and brigadier general, respectively, are reserved for the artillery, engineer, and other branches.

[3] The titles of first captain (*primo capitano*) and first lieutenant (*primo tenente*) are given to captains and 1st lieutenants who have held their respective ranks for 12 years.

[4] The *marescialli* from a separate category, and inasmuch as their is no equivalent of this group in the U.S. Army, these titles are not usually translated. A *maresciallo* is not the equivalent of a U.S. warrant officer.

[5] An officer candidate (*aspirante ufficiale*) has a recognized rank.

[6] The corporal (*caporale*) in the Italian Army is not a noncommissioned officer, but is classed with the private, first class (*appuntato*), both grades being called *graduate*.

Table 10.— Italian military ranks (continued).

Grade	Royal Italian Air Force (*Regia Aeronautica*)	Royal Italian Navy (*Regio Marina*)
Commander-in-Chief	Maresciallo dell'Aira	Grande Ammiraglio
General	Generale d'Armata	Ammiraglio d'Armata
Lieutenant General	Generale designate d'Armata	Ammiraglio designato d'Armata
Major General	Generale di Squadra	Ammiraglio di Squadra
Brigadier General	Generale de Brigata	Ammiraglio di Divisione
Colonel	Colonello	Contrammiraglio
Lieutenant Colonel	Tenente Colonnello	Capitano di Vasecello
Major	Maggiore	Capitano di Fregata
Captain	Capitano[1]	Capitano di Corvetta
1st Lieutenant	Tenente[1]	Tenente
2nd Lieutenant	Sottotenente	Guardiamarina
Officer candidate	Aspirante[4]	—
(no U.S. equivalent)	Maresciallo di 1e Classe[2] Maresciallo di 2e Classe Maresciallo di 3e Classe	Capo di 1e Classe Capo di 2e Classe Capo di 3e Classe
Sergeant-major	Sergente maggiore	Secondo capo
Sergeant	Sergente	Sergente
Lance-corporal	Aviere Scelto	Sottocapo
Private, first class	—	Comuna di 1e Classe
Private	Aviere	Comuna di 2e Classe

Notes:

[1] The titles of first captain (*primo capitano*) and first lieutenant (*primo tenente*) are given to captains and 1st lieutenants who have held their respective ranks for 12 years.

[2] The *marescialli* from a separate category, and inasmuch as their is no equivalent of this group in the U.S. Army, these titles are not usually translated. A *maresciallo* is not the equivalent of a U.S. warrant officer.

A GLOSSARY OF WORLD WAR II MILITARY TERMS

Table 11.— Japanese military ranks.

Grade	Imperial Japanese Army & Navy	Normal command
Field Marshal	Gensui[1]	—
General	Taishō	Army commander
Lieutenant General	Chūjō	Division commander
Major General	Shōshō	Infantry group or brigade commander
Colonel	Tiasa	Regiment commander
Lieutenant Colonel	Chūsa	Regiment second-in-command
Major	Shōsa	Battalion commander
Captain	Tai-i	Company commander
1st Lieutenant	Chū-i	Platoon commander
2nd Lieutenant	Shō-i	Platoon commander
Warrant officer	Jun-i	Command and administrative duties
Sergeant-major	Sōchō	First sergeant
Sergeant	Gunsō	Squad (section) leader
Corporal	Gochō	Squad (section) leader
Lance corporal	Heichō	
Superior private	Jōtōhei	
Private, 1st class	Ittōhei	
Private, 2nd class	Nitōhei	

Note.—The Japanese had no brigadier general rank.

[1] An honorary rank granted by the Emperor to generals.

MILITARY RANKS

Table 12.— Soviet military ranks.

U.S. Equivalent	Army and Air Forces	Navy
Commander in Chief	Generalissimus[1]	—
General of the Army	Marshal Sovetskovo Soyuza	—
(No comparable rank)	Glavnyi Marshal (of an arm)[2]	—
(No comparable rank)	Marshal (of an arm)[2]	—
General	General Armii[2]	Admiral Flota
Lieutenant General	General Polkovnik	Admiral
Major General	General Leitenant	Vitse-Admiral
Brigadier General	General Maior	Kontr-Admiral
Colonel	Polkovnik	Kapitan 1-go ranga
Lieutenant Colonel	Podpolkovnik	Kapitan 2-go ranga
Major	Maior	Kapitan 3-go ranga
Captain	Kapitan	Kapitan-leitenant
(No comparable rank)	Starshii Leitenant	Starshii Leitenant
First Lieutenant	Leitenant	Leitenant
Second Lieutenant	Mladshii Leitenant	Mladshii Leitenant
Officer cadet	Kursant	Kursant
Warrant Officer	—	Michman
Master or First Sergeant	Starshina	—
Staff Sergeant	Starshiii Sershant	Glavny starshina
Sergeant	Sershant	Starshina 1 statie
Corporal	Mladshii Sershant	Starshina 2 statie
Private first class	Yefrejtor	Starshii Krasnoflotz
Private	Krasnoarmeets	Krasnoflotez

Notes.:

[1] Title held by Joesph Stalin, Chairman of the Politboro of the Central Committee (the Communist Party.

[2] The rank of General (*General Armii*) was only awarded to officers of the army fro(the infantry branch directly, all other branches and services were promoted to th(ranks of Marshal and Chief Marshal (*Glavnyi Marshal*) of the branch.

ABBREVIATIONS USED IN THE U.S. ARMY

The following list of authorized abbreviations was compiled from Basic Field Manual 21-30 and Army Regulations 850-150.

A	Army	AEF	American Expeditionary Forces
AA	Antiaircarft		
AAA	Antiaircraft artillery		
AAF	Army Air Force	A Eng Serv	Army Engineer Service
AAIS	Antiaircraft Artillery Intelligence Ser-	A Ex Serv	Army Exchange Service
AB	Air Base		
Abn	Airborne	AF	Air Force
AC of S	Assistant Chief of Staff	AFCC	Air Force Combat Command
AC of S, G–1	Assistant Chief of Staff, Personnel Division	AG	Adjutant General
		AGD	Adjutant General's Department
AC of S, G–2	Assistant Chief of Staff, Military Intelligence Division	ADG	Army Ground Forces
		AGO	Adjutant General's Office
AC of S, G–3	Assistant Chief of Staff, Organization and Training Division	AHQ	Army Headquarters
		AIC	Army Industrial College
		A Int	Air Intelligence
AC of S, G–4	Assistant Chief of Staff, Supply Division	AIS	Artillery Information Service
		Almt	Allotment
AC of S, OPD	Assistant Chief of Staff, War Plans Division	Alot	Alloted
		ALP	Ambulance loading post
Actg	Acting	Alws	Allowancee
AD	Active Duty	AM	Ante meridian (before noon)
ADC	Aide-de-camp		
A Dist	Air District	Am	Ammunition
Adj	Adjutant	AM 1 cl	Air Mechanic, first class
Adm	Administration		
Adm Ser	Administrative Services	AM 2 cl	Air Mechanic, second class
ADTELP*	Advise by teletype	Amb	Ambulance
Adv	Advance	AMC	Army Medical Cops
Adv Gd	Advance Guard	AMecz	Antimechanization or Antimechanized
Adv Msg Cen	Advance Message Center	AMP	Army Mine Planter
		Amph Comd	Amphibious Command

227

AMPS	Army Mine Planter Service	Aux	Auxiliary
		Av	Average
		Avn	Aviation
Am Tn	Ammunition Train	AW	Aircraft warning
ANC	Army Nurse Corps	AW	Article of War
Anl	Animal	AWC	Army War College
Anl-d	Animal-drawn	AWOL	Absent without leave
Ap	Airplane		
Apmt	Appointment	Ax Sig Com	Axis or axes of signal communication
APO	Army Post Office		
App	Appendix *or* Apprehension	BA	Branch assignment
		Bag	Baggage
Apr	April	BAR	Browning automatic rifle
APS	Army Postal Service		
Aptd	Appointed	Bar Bln	Barrage balloon
AR	Army Regulations	BB	Bureau of the Budget
Ar	Arrest		
A/r	At the rate of	BBTC	Balloon Barrage Training Center
ARC	American National Red Cross		
		Bcl	Bicycle
Armd-C	Armored car	Bclt	Bicyclist
Armd F	Armored Force	Bd	Boundary
Armr	Armorer	Bdg	Bridge
Ars	Arsenal	Bglr	Bugler
Artif	Artificer	BHQ	Brigade Headquarters
Arty	Artillery		
Asgd	Assigned	Bkry	Bakery
Asgmt	Assignment	Bks	Barracks
Ash	Airship	Blksm	Blacksmith
Asst	Assistant	Bln	Balloon
Asst Sec War	Assistant Secretary of War	BM	Bench mark
		Bomb	Bombardment
Asst Sec War (Air)	Assistant Secretary of War for Air	BOQ	Bachelor officers' quarters
AT	Antitank	B & Q	Barracks and quarters
ATC	Air Transport Command		
		Br	Branch
Atchd	Attached	B Ren	Battle reconnaissance
Atk	Attack		
ATS	Army Transport Service	Brig	Brigade
		Brig Gen	Brigadier General
Aug	August	Br Tn	Bridge train
Auth	Authorizes	Bsc	Basic
Auto	Automatic	BSTrk	Bomb service truck
AUS	Army of the United States	Bty	Battery
		Btry	Battery

Bul	Bulletin	Clr	Colored
C	Changes *or* Chief *or* Combat *or* Corps	CM	Court martial
		cm	Centimeter
CA	Coast Artillery	Cml	Chemical
CAC	Coast Artillery Corps	Cml Mort	Chemical mortar
cal	Caliber	CMP	Corps of Military Police
Cam	Camouflage		
Capt	Captain	Cmpn	Compilation
Carr	Carrier	CMTC	Citizens' Military Training Camp
Cas	Casualty		
C Auth	Civil Authorities	CO	Commanding Officer
Cav	Cavalry	Co	Company
Cav DHQ	Cavalry division headquarters	C of	Chief of
		C of A	Chief of Artillery
CD	Coast defense	C of Avn	Chief of Aviation
CDD	Certificate of Disability for Discharge	C of S	Chief of Staff
		Col	Colonel
		Colg	College
CE	Corps of Engineers	Coll	Collecting
Cem	Cemetery	Com	Communication
Cen	Center	Co M	Company, motor
Cen	Central	Comd	Command
Cert	Certificate	Comdg	Commanding
CF	Coastal Frontier	Comdr	Commander
Cfr	Chauffeur	Comdt	Commandant
CG	Commanding General	Coml	Commercial
		Comm	Commissary
C & GS Sch	Command and General Staff School	Comtn	Commutation
		Com Z	Communications zone
Ch	Chaplain		
Ch	Church	Conc	Concentration
Char	Character	Conf	Confined
CHQ	Corps headquarters	Conf	Confinement
C in C	Commander in Chief	Confer	Conference
Cir	Circular	Con objtr	Conscientious objector
Civ	Civil *or* Civilian		
Ck	Cook	Con Py	Contact pary
CL	Close-in	Cons	Construction
Cl	Class	Contr	Contract
Cl I Sup	Class I Supplies	Contd	Continued
Cld	Colored	Contl	Control
Cler	Clerk	Conv	Convalescent
Clk	Clerk	Convn	Convenience
clm	Column	CP	Command Post
Clo	Clothing	Cpl	Corporal
Clr	Clearing		

A GLOSSARY OF WORLD WAR II MILITARY TERMS

CPX	Command post exercise	Div OO	Division ordnance officer
CR	Crossroads	DO	Duty officer
cs	Current series	Doc	Document
C Sig O	Chief Signal Officer	DOL	Detached Officers' List
C Tn	Combat Train	DP	By direction of the President *or* Distributing Point
Cur	Current		
CW	Continuous wave		
CWO	Chief warrant officer		
CWS	Chemical Warfare Service	Dp	Dump
		Dpt	Deposit
Cyl	Cylinder	Dr	Drawn
DC	Dental Corps	Drftm	Draftsman
DC of S	Deputy Chief of Staff	Drpd	Dropped
Dec	December	Ds	Distant
Decon	Decontamination	DS	Distant Surveillance *or* Detached Service
Def	Defense		
Del acct	Delinquent account		
DEO	District engineer officer	DSC	Distinguished Service Cross
Delv	Delivered	DSM	Distinguished Service Medal
Delv	Delivery		
DEML	Detached Enlisted Men's List	Dsmtd	Dismounted
		Dy	Duty
Dent	Dental	E	East
Dep	Depot	EAD	Extended active duty
Dept	Department		
Des	Deserted *or* Desertion	Ech	Echelon
		EG	Expert Gunner
Det	Detachment	Elec	Electrical *or* Electrician
DFC	Distinguished Flying Cross		
		Elim	Eliminate
DHQ	Division Headquarters	Elm	Element
		EM	Enlisted man or men
Dir	Director	Emb	Embrakation
Dis	Disciplinary	Emrg	Emergency
Disab	Disability	Engr	Engineer
Disch	Discharge *or* Discharged	Engrs	Engineers
		Enl	Enlisted
Discontd	Discontinued	Enlmt	Enlistment
Dishon	Dishonorable *or* Dishonorably	EP	Entrucking point
		ER	Expert Rifleman
Dismd	Dismissed	ERC	Enlisted Reserve Corps
Dist	District		
Div	Division	ETS	Expiration Term of Service

Evac	Evacuation	Fy	Fiscal year
Evac Hosp	Evacuation hospital	G	Gun
Evid	Evidence	G–1	General Staff first
Ex	Excellent or Execu-		section or As-
	tive		sistant chief of
Excl	Exlusive		staff for personnel
EXFORACT*	Extracted for action	G–2	General Staff second
Exp	Expiration		section or As-
EXREQ*	Extract of requisi-		sistant chief of
	tion		staff for military
F	Field		intelligence
FA	Field Artillery	G–3	General Staff third
FD	Finance Departe-		section or As-
	ment		sistant chief of
Feb	February		staff for opera-
Fed	Federal		tions and training
Fi	Fighter	G–4	General Staff fourth
Fil	Filter		section or As-
Fil Cen	Filter center		sistant chief of
Fin	Finance		staff for supply
Fin O	Finance officer	Gar	Garage
1st	First	Gas NCO	Gas noncommis-
1 Cl	First class		sioned officer
1st Lt	First lieutenant		
Fis	Fiscal	Gas O	Gas officer
Fl	Flash	GSM	General court mar-
F Lab	Field laboratory		tial
Fld	Field	Gd	Guard
Flt	Flight	Gen	General
FM	Field Manual	Gen Disp	General dispensary
FO	Field Orders	Gen Hosp	General hospital
Forf	Forfeit	Gen Serv	General Service
FR	Flash ranging	Geol	Geological
fr	From	GHQ	General Headquar-
Fraud	Fraudulant		ters
FS	Final Statement	GHQ AF	General Headquar-
FS	Film strip		ters Air Force
FSR	Field Service Regu-		
	lations	GN*	Army Ground Forces
ft	Feet	G–NP	Chemical agent,
Ft	Fortification		nonperistant
F Tn	Field Train	Gnr	Gunner
Fur	Furlough	GO	General Orders
Furn	Furnished	Govt	Government
Fwd Ech	Forward echelon	G–P	Chemical agent,
Fxd	Fixed		persisant

Gp	Group	I & I Report	Inventory and Inspection Report
G-PF	Gasproof dugout or building	Impreg	Impreginating
Gpmt	Groupment	in	Inch
G Reg	Graves registration	Incl	Inclosure
Grad	Graduate	incl	Inclusive
GS	General Staff	Incld	Included
GSC	General Staff Corps	Inctd	Inducted
GSS	General Staff Service Schools	Ind	Indorsement
		Inf	Infantry
H	Heavy or Horse	Info	Information
Har	Harbor	Ins	Insurance
HD	Harbor defense	Insp	Inspector
HDC	Harbor defense command	Instl	Installation
		Instr	Instrument
H-Dr	Horse-drawn	Int	Intelligence
HE	High explosive	Intn	International
H & Mecz	Horse and mechanized	Int O	Intelligence officer
		Intpr	Interpreter
Hon	Honor or Honorable	IP	Initial point
Hosp	Hospital	JA	Judge Advocate
How	Howitzer	JAG	Judge Advocate General
Hq	Headquarters		
Hq Comdt	Headquarters commandant	JAGD	Judge Advocate General's Department
Hq Comdt & PM	Headquarters commandant and provost marshal		
		Jan	January
		Jd	Joined
Hq & Hq Co	Headquarters and headquarters company	Jn	Join
		Jr	Junior
		Jul	July
Hq & Hq Det	Headquarters and headquarters detachment	Kd	Killed
		Ki	Kitchen
		L	Light
Hq & Serv Co	Headquarters and service company	Lab	Laboratory
		Lbr	Labor
H & RP	Holding and reconsignment point	LC	Line of communication
Hv	Heavy	LD	Line of duty or Line of departure
Hv W	Heavy weapons		
Hwy	Highway	Ldr	Leader
I	Interceptor	Ldry	Laundry
IC	Information center	LDS	Line of duty status
IG	Inspector General	LE	Low explosive
IGD	Inspector General's Department	Leg	Legislative
		Lgts	Lights

LM	Land mine	Mil av	Military aviator
LMG	Light machine gun	Misc	Miscellaneous
Lmn	Lineman	Mk	Mark
Ln	Liaison	MLR	Main line of resis-tance
Ln O	Liaison Officer		
LP	Livens projector	Mm	Marksman
LR	Long range	mm	Millimeter
Lt	Lieutenant	Mort	Mortar
Lt Col	Lieutenant Colonel	mos	Months
Ltd serv	Limited service	MP	Military Police
Lt Gen	Lieutenant General	Mph	Miles per hour
Ltr	Letter	Mr	Master
Lv	Leave	MR	Medium range *or* Mobilization Regu-lations.
M	Motor *or* Medium		
Machst	Machinist		
Maint	Maintenance	M/R	Memorandum Re-ceipt *or* Morning report
Maint of E	Maintenance of equipment		
Main Py	Maintenance party	M & S	Maintenance and supply
Maj	Major		
Maj Gen	Major General	Msg	Message
Man	Manual	Msg cen	Message Center
Mar	March (month or movement)	Msg DPU	Message dropping and pickup ground
Mat	Material	M sgt	Master sergeant
Mbl	Mobile	MSR	Main supply road
MC	Medical Corps	MT	Motor Transport
MCM	Manual for Courts-Martial	Mtcl	Motorcycle
		Mtclt	Motorcyclist
MD	Medical Department	Mtricl	Motor-tricycle
Mec	Mechanic	MTP	Mobilization Train-ing Program
Mdl	Model		
Mech	Mechanic	Mtz	Motorized
Mecz	Mechanized	Mun	Munitions
Med	Medical	Mun O	Munitions Officer
Med Adm C	Medical Administra-tive Corps	Mus	Musician
		N	North
		Nav	Navigation *or* Naval *or* Navy
Memo	Memorandum		
Met	Meterorological	NGB	National Guard Bu-reau
MG	Machine gun		
MH	Medal of Honor	NCO	Noncommissioned officer
MI	Military Intelligence		
mi	Mile	NCS	Net Control Station
Mil	Military	No	Number
Mill att	Military Attaché	Nov	November

Nt	Night	P & D Sec	Pioneer and demoli-tion section
NY	New York		
O	Office *or* Officer or Order(s)	P/E	Port of embarkation
		Pers	Personnel
O/a	On or about	Perm	Permanent
Obsn	Observation	Pfc	Private, first class (*see also* Pvt 1cl)
Obsr	Observer		
OC	Officer candidate *or* Officer in charge	Pgn	Pigeon
		Pgn Co	Pigeon company
OCNGA	Officer in charge of National Guard affairs	PH	Purple Heart
		P Hosp	Post hospital
		Photo	Photograph *or* Pho-tographic
OCS	Officer candidate school		
		Pion	Pioneer
Oct	October	Pk	Park
OD	Officer of thedya	Pkmr	Packmaster
od	Olive drab	Pkr	Packer
Odly	Orderly	PL	Post Laundry
OG	Officer of the guard	Plat	Platoon
1 Pdr	One pounder	PM	Post meridian (after noon) *or* Provost Marshal
OO	Ordnance officer		
OP	Observation point		
OPD	Operations Division	PMG	Provost Marshal General
OPL	Outpost line		
OPLR	Outpost line of resis-tance	PMS & T	Professor Military Science and Tac-
Opn	Operation	Pmt	Payment
Opr	Operator	Pn	Patient
Ops	Opinions	Pon	Pontoon
OQMG	Office of the Quar-termaster General	Post	Postal
		P/R	Pay roll
		Prcht	Parachute
OR	Organized Reserve	Prim	Primary
Ord	Ordnance	Princ	Principal
Ord Dpt	Ordnance Depart-ment	Pris	Prisoner
		Prk	Park
Orgn	Organization	Prof	Professor
O S & D R	Over, Short, and Damaged Report	Prov	Provisional
		PS	Philippine Scout
OU Sec War	Office Under Secre-tary of War	Pt	Point (*see also* P)
		Pub	Public
P	Point (*see also* Pt)	Publ	Publication
Par	Paragraph	Pur	Pursuit
Part	Partial	Pvt	Private
Pd	Paid	Pvt 1cl	Private, first class (*see also* Pfc)

PW	Prisoner of war	Rejd	Rejoined
Py	Party	Reld	Relieved
QM	Quartermaster	REPHONE*	Reference telephone conversation
QMC	Quartermaster Corps	Repl	Replacement
QMG	Quartermaster General	Repr	Reproduction
		Req	Requisition
QMSO	Quartermaster Supply Officer	Reqmt	Requirements
		Requal	Requalified
Qr	Quartering	Res	Reserve
Qrs	Quarters	Resc	Rescinded
Qual	Qualified	Resgd	Resigned
R	Regulating or Rifle	Ret	Retired
RA	Regular Army	Retmt	Retirement
Rad	Radio	Rets	Returns
RADAR	Radio detector equipment	Rg	Range
		Rhd	Railhead
Rad Int	Radio intelliegence	RHO	Railhead officer
RAR	Regular Army Reserve	RHQ	Regimental headquarters
Rat	Rations	RJ	Road junction
RB	Road bend	Rmt	Remount
Rcd	Record	RO	Regimental Orders
Rcn	Reconnaissance	ROTC	Reserve Officers' Training Corps
Res	Resources		
Rct	Recruit	Rmp	Revolutions per minute
Rctg	Recruiting		
Rd	Reduce or Reduced or Road	R & P Sec	Radio and panel section
Rdg	Reducing	Rpt	Report
RDP	Ration distributing point	RR	Railroad
		Rr	Rear
REACK*	Receipt acknowledged	RRL	Regimental reserve line
Reaptd	Reappointed	RS	Road space
Rec	Recreation	R/S	Report of Survey
Recd	Received	R sta	Regulating station
Recons Prk	Reconstruction park	RTC	Replacement Training Corps
Recp cen	Reception center		
Reenl	Reenlist	RTO	Railroad transportation officer
Reenlmt	Reenlistment		
Ref	Refrigeration		
Reg	Regiment	R Traf O	Railway traffic officer
Regt	Regiment		
Regtl	Regimental	Ry	Railway
Rein	Reinforced	S	Service

S	Sharpshooter	Serv C	Service command
S–1	Battalion (or regi-mental) adjutant	S & F	Sound and flash
		SG	Surgeon General
S–2	Battalion (or brigade or regi-mental) intelli-gence officer	SGO	Surgeon General's Office
		Sgt	Sergeant
		SHIPIM*	Ship immediately
		Shipt	Shipment
S–3	Battalion (or brigade or regi-mental) plans and training officer	Sig	Signal
		Sig O	Signal officer
		Sig C	Signal Corps (see also SC)
S–4	Battalion (or brigade or regi-mental) supply officer	Sk	Sick
		SL	Support Line
		S-L	Sound locator
SA	Small arms	SL Btry	Searchlight Battery
Sad	Saddle	SM	Soldier's Medal
S & B	Sterilization and bath	Slt	Searchlight
		SMG	Submachine gun
Sales Comm	Sales commissary	Sn	Sanitary
SAR	Semiautomatic rifle	SO	Special Order
Sb	Switchboard	SOI	Signal operations instruction
SC	Signal Corps (see also Sig C) or Summary Court	SOK*	Supply authorized
		SOP	Standing operating procedure
S/C	Statement of Charges	SOS	Services of Supply
Sc	Sidecar	SP	Supply point
Scd	Scheduled	Sp	Special
Sch	School	Specl	Specialist
SCM	Special Court-Mar-tial	SPM	Self-propelled mount
Sct	Scout	Sp Msgr	Special Messenger
Sct C	Scout car	Sp Trs	Special troops
Sct C Ht	Scout car, half track	Sp Wpn	Special weapons
SD	Special Duty	Sq	Squadron
sd	same date	SR	Sound ranging
Sec	Section	S/R	Service record
2d Lt	Second lieutenant	S Rep	Shoe repair
Sec War	Secretary of War	SS	Selective Service
Sem	Semimobile	S sgt	Staff sergeant
Sent	Sentence	Sta	Station
Sentd	Sentenced	Stab	Stable
Sep	Separate	Sta com	Station complement
Sept	September	Statl	Statistical
Serv	Service	Steno	Stenographer(s)

U.S. ARMY ABBREVIATIONS

Stev	Stevadore	T/E	Tables of equipment
Stew	Steward	Tech	Techncial
Stf	Staff	Techn	Technician
Str	Strength	Telg	Telegram
Strag L	Straggler Line	Temp	Temporary
Stud	Student	TF	Training file
SU	Service unit	Tg	Telegraph
Sub	Subject *or* Submarine	TH	Territory of Hawaii
		THQ	Theater headquarters
Subaq	Subaqueous		
Subm	Submarine	Tk	Tank
Subs	Subsistance	TL	Time Length
Sum	Summary	Tlmkr	Toolmaker
Sup	Supply authorized	Tlr	Trailer
Sup Dep	Supply depot	TM	Technical Manual *or* Trench mortar
Sup Pt	Supply point		
Sup O	Supply officer	Tn	Train
Supp	Supplemental	Tng	Training
Surg	Surgeon *or* Surgical	T/O	Tables of Organization
Surr	Surrender *or* Surrendered		
		T of Opns	Theater of Operations
Surv	Survey		
Susp	Suspended	Top	Topographical
Sw C	Switching central	Topo	Topographic
T	Transport *or* Transportation	Tor	Torpedo
		Tp	Telephone
TA	Territorial Assignement	TPA	Travel by officer or his dependents by privately owned automobile is authorized
T/A	Tables of Allowances		
Tac	Tactial		
TAG	The Adjutant General		
		TR	Technical Regulations *or* Training Regulations
T/BA	Tables of Basic Allowances		
TC	Training circular *or* Transportation Corps	T/R	Transportation request
		Tr	Troop
TD	Tank destroyer	Traf	Traffic
TD	Tractor-drawn (*see also* Tr Dr)	Trac	Tractor
		Tr Dr	Tractor-drawn (*see also* TD)
TDN	Travel directed in necessary in the military service	Trfd	Transferred
		Tricl Mtr	Tricycle, motor
TDT/FC	Tank Destroyer Tactical and Firing Center	Trk	Truck
		Trk Dr	Truck-drawn
		Trk hd	Truck head

Trne	Trainee		Wea	Weather
Trs	Trooops		Wg	Wing
T sgt	Technical sergeant		Wkr	Wrecker
TU	Training unit		Wldr	Welder
TWX	Teletypewriter Ex-		Wn	Winch
	change		WO	Warrant officer
Unasgd	Unassigned		WOJG	Warrant officer, ju-
Unsat	Unsatisfactory			nior grade
USA	United States of		WP	Will proceed to
	America		WPD	War Plans Division
U Sec War	Under Secretary of		Wpn Carr	Weapon carrier
	War		Wrnt	Warrant
USMA	United States Mili-		W Sup	Water supply
	tary Academy		Yd	Yard
USMC	United States Ma-		Z	Zone
	rine Corps		Z of I	Zone of the interior
USN	United States Navy			
USP & DO	United States Prop-			
	erty and Disburs-			
	ing Officer			

USN — United States Navy

USP & DO — United States Property and Disbursing Officer

Note: Abbreviations marked with asterisks indicate they were used in telegraphic and radio communications.

VC	Veterinary Corps
Vet	Veterinary
VO	Verbal Order
VOC	Volunteer officer candidate
Vol	Volunteer
Vou	Voucher
W	West
WAAC	Women's Army Auxiliary Corps
Wag	Wagon
WD	War Department
WD*	War Department General Staff
WDCSA*	Chief of Staff, U. S. Army
WDGAP*	Assistant Chief of Staff, G–1
WDGBI*	Assistant Chief of Staff, G–2
WDGCT*	Assistant Chief of Staff, G–3
WDGDS*	Assistant Chief of Staff, G–4
WDOPD*	Assistant Chief of Staff, OPD

ABBREVIATIONS USED IN THE BRITISH ARMY

The following list of authorized abbreviations was compiled from Field Service Pock Book, Part I—Pamphlet No. 3, Abbreviations, 1944.

AA	Anti-aircraft *or* Army Act	AOD	Advanced ordnance depot
AAD	Advanced ammunition depot	AP	Ammunition point *or* armour piercing
AB	Army book	A per	Anti-personnel
accn	Accommondation	APIS	Army photographic interpretation sec-
ACI	Army Council Instruction	APO	Army post office
ACIGS	Assistant Chief of the Imperial General	approx	Approximately *or* approximate
ack	Acknowledge, acknowledged, *or* acknowledgment	appx	Appendix
		armd	Armoured
		armd C	Armoured car
ACV	Armoured combat vehicle	ARG	Armoured replacement group
addsd	Addressed	ARH	Ammunition railhead
adjt	Adjutant	ARO	Army routine order
adm	Administration *or* administrative	ARP	Air raid precautions *or* ammunition refilling point
ADC	Aide-de-camp		
ADS	Advanced dressing station	arty	Artillery
		arty R	Artillery reconnaissance
adv	Advance *or* advanced		
AF	Army form	art wks coy	Artisan works company
AFV	Armoured fighting vehicle	ASSU	Air support signal unit
AG	Anti-gas	ASD	Ammunition sub de-
AGRA	Army group, Royal Artillery	aslt	Assault
		AT	Animal transport
AGRE	Army group, Royal Engineers	A tk	Anti-tank
		ATI	Army Training Instruction
airtps	Airborne troops		
ALG	Advanced landing ground	ATM	Army Training Memorrandum
ALO	Air liaison officer	A tps	Army Troops
AL sec	Air liaison section	att	Attack, attached, *or* attachment
amb	Ambulance		
amn	Ammunition	Aust	Australia *or* Australina
annx	Annexure	AVRE	Assault vehicle, Royal Engineers
AO	Army Order		
AOC	Air officer command-		

A GLOSSARY OF WORLD WAR II MILITARY TERMS

BAD	Base ammunition depot	CDI	Command driver increment
BBP	Bulk breaking point	Cdn	Canadian
BC	Battery commander	cdo	Commando
BD	Base depot *or* bomb disposal	CF	Chaplain to the forces
		cfn	Craftsman *or* crafts-
bde	Brigade	CGS	Chief of the General Staff
bdr	Bombadier		
bdy	Boundary	chem	Chemical
BGS	Brigadier, General	C in C	Commander-in-Chief
BM	Brigade Major (General Staff)	CIGS	Chief of the Imperial General Staff
BMA	Beach maintenance area	civ	Civil *or* civilian
		CL	Centre line
bn	Battalion	COO	Chief ordnance officer
BOD	Base ordnance depot	col	Colonial
bomrep	Bombing report	coln	Column
BOWO	Brigade ordnance warrant officer	comd	Command, commanded, commandant, *or* com-mander
BQMS	Battery quarter-master-general		
		CO	Commanding officer
br	Bridge *or* bridging	comn	Communication
Brig	Brigadier	comp	Composite
Brit	British	conc	Concentrate, concentrated, *or* concentra-toin
BSD	Base supply depot		
BSM	Battery serjeant-major		
B sup airfd	Base supply airfield	con R	Contact reconnais-
bty	Battery	constr	Construct, constructed, *or* construction
BU	Base unit		
B wksp	Base workshop	coord	Co-ordinate, co-ordinated, co-ordinating, *or* co-ordination
CA	Civil affairs *or* coast artillery		
cam	Camouflage *or* camouflaged	coy	Company
		cpl	Corporal
capt	Captain	CQMS	Company quarter-master serjeant
cas	Casualty(ies)		
cav	Cavalry	CS	Close support
CB	Confinement to barracks *or* counter-battery	CSM	Company serjeant-major
		C tps	Corps troops
CBO	Counter-battery offi-	CV	Lorry, command
CCP	Casualty collecting	CW	Chemical warfare
CCS	Casualty clearing station	cwt	Hundredweight(s)
		DCIGS	Deputy Chief of the Imperial General
CD	Coast defence		
Cda	Canada	decn	Decontamination

def	Defend, defended, defence, *or* defensive
del	Deliver, delivered, *or* delivery
dep	Depot
derv fuel	Diesel engined road vehicle fuel
det	Detach, detached, *or* detachment
DF	Defensive fire *or* direction finding
DID	Detail issue depot
dis P	Dispersal point
dist	District
div	Division *or* divisional
DP	Delivery point
DR	Despatch rider *or* motor-cyclist
DRLS	Despatch rider letter service
DS	Dressing station
dvr	Driver
DZ	Dropping zone
EA	East Africa, East African, *or* enemy air-
ech	Echelon
EF	Expeditionary force
eg	For example
EME	Electrical mechanical engineer
emp	Employ, employed, *or* employment
engr	Engineer
eqpt	Equipment
ESO	Embarkation staff officer
est	Establish, established, *or* establishment
etc	And so on *or* and the rest
evac	Evacuate, evacuated, *or* evacuation
excl	Exclude, excluded, excluding, *or* exclu-
FBE	Folding boat equipment

fd	Field
fd bchy	Field butchery and cold storage depot
fd bdy	Field bakery
FDL	Forward defended
FDS	Field dressing station
FF	Field force or fire fighting
FGCM	Field general court martial
flt	flight
FMA	Forward maintenance area
FOO	Forward observation officer
freq	Frequency
FS	Field service *or* field security
F sp	Flash spotting
FSPB	Field service pocket book
FSS	Fixed signal service
FSU	Field surgical unit
ft	Foot *or* feet
FTU	Field transfusion unit
fmn	Formation
FUP	Forming up place
fus	Fusilier(s)
fwd	Forward *or* forwarded
gal	Gallon
GCM	General court-martial
gd	Guard
gdsm	Guardsman *or* guardsmen
gen	General
GHQ	General Headquarters
gnr	Gunner
GOC (in C)	General Officer Commanding (-in-Chief)
gp	Group
grn	Garrison
GRO	General routine order
GS	General service, general staff, *or* General Staff Branch

GSO	General Staff Officer
GT coy	General transport company
HAA	Heavy Anti-Aircraft
HD	Home defence or horse drawn
HE	High explosive
HF	Harassing fire or high frequency
HG	Home Guard
hosp	Hospital
how	Howitzer
HP	High power or horse power
HQ	Headquarter(s)
hr	Hour
HT	High tension or horsed transport
hy	Heavy
hyg	Hygiene
IA	Indian Army
IC	In charge or internal combustion
ie	Namely or that is to
in	Inch
incl	Include, included, including, or inclusive
Ind	India or Indian
indep	Independent
inf	Infantry
infm	Inform, informed, or information
instr	Instruct, instructed, instruction, or instructor
int	Intelligence
INT	Intelligence (General Staff)
intercomm	Intercommunication
IO	Intelligence officer
IOO	Inspecting ordnance officer
IS	Internal security
IWT	Inland water transport
LAA	Light Anti-Aircraft
lab	Laboratory or labour

LAD	Light aid detachment
lb	Pound(s)
L bdr	Lance-bombadier
L cpl	Lance-corporal
ldg	Landing
LMG	Light machine gun
LO	Liaison officer
L of C	Line, or lines, of communication
lor	Lorried or lorryborne
L sjt	Lance-serjeant
lt	Lieutenant or light
LT	Line telegraphy or low tension
LZ	Landing zone
MA	Military assistant or Military Attaché
MAC	Motor ambulance company
mag	Magazine or magnetic
maint	Maintain, maintained, or maintenance
maj	Major
MC	Motor-cycle or movement control
mech	Mechanic, mechanical, or mechanized
med	Medium or medical
met	Meteorological or meteorology
MFO	Military forwarding officer
MFPS	Mobile field photographic section
MG	Machine gun
MGGS	Major General, General Staff
mih	Miles in the hour
mih2	Miles in two hours
mil	Military
min	Minute
mk	Mark
MLBU	Mobile laundry and bath unit
MMG	Medium machine gun
MO	Medical officer

mob	Mobile *or* mobilization	pl	Platoon
mot	Motor *or* motorized	PM	Provost Marshal
mov	Mouvement	pmr	Paymaster
MP	Meeting point *or* military police	pnr	Pioneer
		PO	Post office
mph	Miles per hour	POL	Petrol, oil, and lubricants
MS	Military secretary		
msg	Message	posn	Position
MT	Mechanical transport	PP	Petrol point
mtd	Mounted	PRH	Petrol railhead
mtn	Mountain	pr	Pounder
MTP	Military Training Pamphlet	pro	Provost
		PRP	Petrol refilling station
NCO	Non-commissioned officer	pt	Point
		pte	Private
Nfld	Newfoundland	PW	Prisoner(s) of war
NZ	New Zealand	QM	Quarter-master
OBD	Ordnance beach detachment	QMS	Quarter-master-serjeant
OC	Officer commanding	RAP	Regimental aid post
offr	Officer	rd	Road
OIC	Officer-in-charge (of)	RDI	Relief driver incre-
OO	Operation order *or* ordnance officer	rec	Recover, recovered, or recovery
op	Operate, operated, operation, operational, *or* operator	recce	Reconnaissance *or* reconnoitre
		ref	Reference
OP	Observation post	reg	Regulate, regulated, regulating, *or* regulation
OPS	Operations (General Staff)		
OR	Other rank(s)	regt	Regiment *or* regimen-
ord	Ordnance	rep	Represent, represented, *or* representative
org	Organize, organised, *or* organizational		
		res	Reserve(s)
oz	Ounce(s)	rfn	Rifleman *or* riflemen
PA	Personal Assistant	rft	Reinforcement
PAD	Passive Air Defence	rft HU	Reinforcement holding unit
para	Parachute *or* para-		
pet	Petrol (gasoline)	rft SU	Reinforcement sub-
ph	Photograph *or* photographic	RH	Railhead
		RHA	Royal Horse Artillery
photogrid	Photograph, vertical air, gridded	RHMA	Railhead, *or* roadhead, maintenance area
ph R	Photographic reconnaissance	rly	Railway
		RMA	Rear maintenance
pk	Park	RO	Routine order

ROO	Railhead, *or* roadhead, ordnance officer	SP	Self-propelled *or* starting point
RP	Refilling point, regimental police, *or* Rules of Procedure	spr	Sapper
		SQMS	Staff, *or* squadron, quarter-master-serjeant
rpg (pm)	Rounds per gun (per minute)	sqn	Squadron
rptd	Repeated	SRH	Supply railhead
RQMS	Regimental quarter-master-serjeant	S rg	Sound ranging
		S sjt	Staff serjeant
RSD	Returned stores depot	SSM	Staff, <u>or</u> squadron, serjeant-major
RSM	Regimental sergeant-major		
R sup O	Railhead, *or* roadhead, supply officer	S sup O	Senior supply officer
		stereo	Stereoscope *or* sterographic
RT	Radio telephony	str	Seater *or* strength
RTO	Railway traffic officer	strat R	Strategical reconnaissance
RV	Rendezvous		
RW	Royal Warrant for Pay and Promotion	sup	Supply *or* supplied
		sup P	Supply pont
SA	Small arms, South Africa, *or* South African	svy	Survey
		swbd	Switchboard
SAA	Small arms ammunition	TA	Territorial Army
		tac R	Tactical reconnais-
sal	Salvage *or* salvaged	tech	Technical
SB	Stretcher bearer	TC	Traffic control
SBG	Small box girder	TCV	Troop carrying vehicle
SD	Staff Duties (and Training), General	tcl	Tentacle
		tele	Telephone
SDP	Supply dropping point	tfc	Traffic
sec	Second *or* section	tg	Telegraph *or* telegraphic
2IC	Second in command		
shelrep	Shelling report	tk	Tank
sig	Signal	TMC	Thompson machine carbine
sigmn	Signalman *or* signalmen		
		tn	Transportation
sitrep	Situation report	TO	Transport officer (*only when used with bde, bn, etc.*)
sjt	Serjeant		
SL	Searchlight *or* start		
SLA	Supply loading airfield	TP	Traffic post
SLP	Supply landing point	tp	Troop
SMC	Sten machine carbine	tpr	Trooper
SMO	Senior medical officer	tpt	Transport
SO	Staff officer	tptd	Transported
sp	Support *or* supported	trg	Training
		tptr	Transporter

BRITISH ARMY ABBREVIATIONS

UK	United Kingdom
UM	Urgent message
USA	United States of America
UXB	Unexploded bomb(s)
VCIGS	Vice Chief of the Imperial General Staff
veh	Vehicle
vet	Veterinary
VP	Vulnerable point
VRD	Vehicle reserve depot
vtm	Vehicles to the mile
WA	West Africa *or* West African
WE	War establishment
wef	With effort from
wh	Wheel *or* wheeled
wksp	Workspho
WL	Wagon line
WO	Warrant officer
WP	Water point
wpfg	Waterproofing
wrls	Wireless
WS	War substantive, warlike store, *or* wireless set
WT	Weapon training *or* wireless telegraphy
WWCP	Walking wounded collecting point
X rds	Crossroads
yd	Yard

LIST OF REFERENCES

U.S. War Department Documents

AR 850–150, Army Regulations, Authorized Abbreviations for Military Records, 1943.

FM 1-5, Army Air Forces Field Manual, Employment of Aviation of the Army, 1943.

FM 2–15, Cavalry Field Manual, Employment of Cavalry, 1941.

FM 7–5, Infantry Field Manual, Organization and Tactics of Infantry, Organization and Tactics of the Rifle Battalion, 1940.

FM 10-5, Quartermaster Field Manual, Quartermaster Operations, 1941.

FM 17–10, Armored Force Field Manual, Tactics and Technique, 1942.

FM 21–30, Basic Field Manual, Conventional Signs, Military Symbols, and Abbreviations, 1943.

FM 31–30, Basic Field Manual, Tactics and Technique of Air-borne Troops, 1942.

FM 31-35, Basic Field Manual, Aviation in Support of Ground Forces, 1942.

FM 100-5, Field Service Regulations, Operations, 1941 and 1944.

Military Intelligence Service Special Series No. 13, British Military Terminology, 1943.

TM 20–205, Technical Manual, Dictionary of United States Army Terms, 1944.

TM 30–410, Technical Manual, Handbook on the British Army with Supplements on the Royal Air Force and Civilian Defense Organizations, 1942.

TM 30-430, Technical Manual, Handbook on U.S.S.R. Military Forces, 1945-46.

TM 30-480, Technical Manual, Handbook on Japanese Military Forces, 1944.

TME 30-420, Technical Manual, Handbook on the Italian Military Forces, 1943.

TM-E 30-451, Technical Manual, Handbook on German Military Forces, 1945.

British War Office Documents

Field Service Pocket Book, 1932.

Field Service Pocket Book, Pamphlet No. 1, Glossary of Military Terms and Organization in the Field, 1940.

Field Service Pocket Book, Part I—Pamphlet No.1, Glossary of Military Terms, 1944.

Field Service Pocket Book, Part I—Pamphlet No. 3, Abbreviations, 1944.

Field Service Regulations, Vol. I, Organization and Administration, 1930, Reprinted with Amendments (Nos. 1-11), 1939.

Field Service Regulations, Vol. II, Operations—General, 1935.

Field Service Regulations, Vol. III, Operations—Higher Formations, 1935.

Handbook of the French Army, 1940.

Handbook of the Russian Army, 1940.

Military Training Pamphlet No. 22, Tactical Handling of Army Tank Battalions, 1939.

Military Training Pamphlet No. 41, The Tactical Handling of the Armoured Division and its Components, 1943.

Military Training Pamphlet No. 63, The Cooperation of Tanks with Infantry Divisions, 1944.

New Notes on the Red Army, No. 2, Uniforms and Insignia, 1944.

The King's Regulations for the Army and the Army Reserve, 1940.

The King's Regulations and Air Council Instructions for the Royal Air Force, 1943.

www.ingramcontent.com/pod-product-compliance
Lightning Source LLC
Chambersburg PA
CBHW060657100426
42735CB00040B/2939